WE:ROBOT

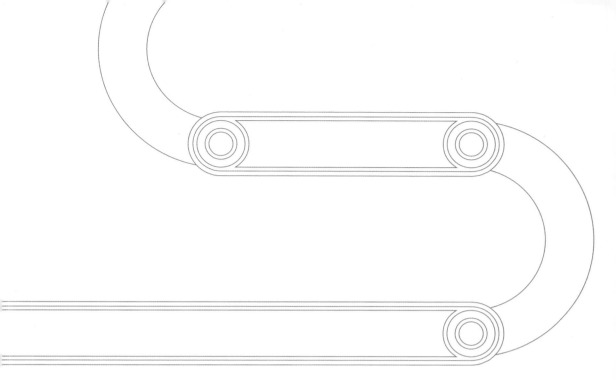

WE:ROBOT

The robots that already rule our world

DAVID HAMBLING

Brimming with creative inspiration, how-to projects and useful information to enrich your everyday life, Quarto Knows is a favourite destination for those pursuing their interests and passions. Visit our site and dig deeper with our books into your area of interest: Quarto Creates, Quarto Cooks, Quarto Homes, Quarto Lives, Quarto Drives, Quarto Explores, Quarto Gifts, or Quarto Kids.

First published in 2018 by Aurum Press
an imprint of The Quarto Group
The Old Brewery, 6 Blundell Street,
London N7 9BH
United Kingdom

www.QuartoKnows.com

A catalogue record for this book is available from the British Library.

ISBN 978 1 78131 746 4
Ebook ISBN 978 1 78131 805 8

10 9 8 7 6 5 4 3 2 1
2022 2021 2020 2019 2018

Design by Ginny Zeal
Illustrations by Liron Gilenberg ironicitalics.com

Printed by RR Donnelley, China

CONTENTS

INTRODUCTION

Robots are changing the world . . . but before we can begin to talk about them, we need to decide what a robot is. What distinguishes a robot from any other machine? The stereotypical robot is a mechanical humanoid that walks and talks like a person, but robots like that only exist – so far – in science fiction. In the real world, most robots neither walk nor talk, and are nothing like people.

Going back to the word's origin is little help. The term 'robot' was first used in Karel Capek's 1921 play R.U.R. for which he created a race of artificial industrial workers – Rossum's Universal Robots – the word 'robot' is derived from a Czech word meaning 'forced labour'. To make matters more confusing, Capek's fictional robots were organic beings that were grown synthetically, more like clones than our idea of robots.

The Oxford English Dictionary suggests that a robot is 'a machine capable of carrying out a complex series of actions automatically'. This would include machines like dishwashers and washing machines, which are – perhaps unfairly – not seen as robots. The International Federation of Robotics defines a robot as 'an automatically controlled, reprogrammable, multipurpose manipulator . . . which may be either fixed in place or mobile'. This may be useful in the industrial field, but misses many others; surgical robots, for example, are not programmable.

Early ideas of robots came from fiction, and developers played up the 'mechanical man' stereotype of literature's invention. 'Elektro the Moto Man' was a metallic human 2m (6.5ft) tall, who gave demonstrations to packed houses at the 1939 New York World's Fair. He could respond to verbal commands, answer questions in a suitably robotic voice, count on his fingers and even smoke a cigarette. Elektro was built by the Westinghouse Electric Corporation to showcase some of its most advanced technology: photocells and electrical relays were cutting-edge industrial technology. Elektro was a fake, his answers pre-scripted and his abilities limited to set tricks, but robotic devices using the same technology were doing real work.

Machines that took over the jobs of humans were often called robots. The first traffic signals were installed outside the Houses of Parliament in 1868, with red, green and amber gas lights for the nights, based on signals that directed trains. A policeman operated them manually. When automated traffic lights were introduced, in the 1920s, they were known as 'robot policemen'.

In the 1940s, Nazi Germany fielded the V-1, a pilotless aircraft carrying explosives. These were known in the English-speaking world as 'robot aircraft' or just 'robots' ('doodlebug' and 'buzz bomb' were less common). We might now call them drones, a term that overlaps confusingly with robots.

For this book, we have accepted a broad definition of what makes a robot. Surgical and bomb-disposal machines that are universally known as robots are included, even though some might argue they are remote-controlled and not 'real' robots. There are plenty of robot vehicles – on the ground, on the water and in the air, as well as underwater and underground. There are also devices that require a human operator, that are unquestionably robotic in nature, such as prosthetic hands or complete exoskeletons. But, perhaps the robots that continue to fascinate us the most are the humanoids.

Serious efforts to build a human-like robot date back at least to designs sketched by Leonardo da Vinci for a mechanical knight in 1495, often cited as the ancestor of modern robots. Leonardo used his knowledge of human anatomy, the workings of joints and muscles to build an artificial human with levers and pulleys. The design, based on a suit of armour, could move its arms and legs and raise its visor. It even had a rudimentary form of programming, as it could be set to perform different actions by changing the settings of gears on a clocklike 'controller'.

We do not know if Leonardo's knight was ever built, but it shows a profound understanding of the issues of robotics, and how problems can be overcome by borrowing techniques from nature. Modern robot arms, with their wrist and elbow joints, bear more than a coincidental resemblance to Leonardo's knight. Given the same challenge of articulation and moving a hand through three-dimensional space, engineers arrive at the same solution.

While robots may have been around for centuries, the current generation offer something entirely new. Now we are moving from the illusion of human capability to the reality. They may walk clumsily, and drive poorly, and lack human skill when it comes to picking up an apple, but robots are learning fast. The day is coming when they will rival and then surpass us in all sorts of fields. These robots are already starting to change the world. In the coming decades, they will do far more.

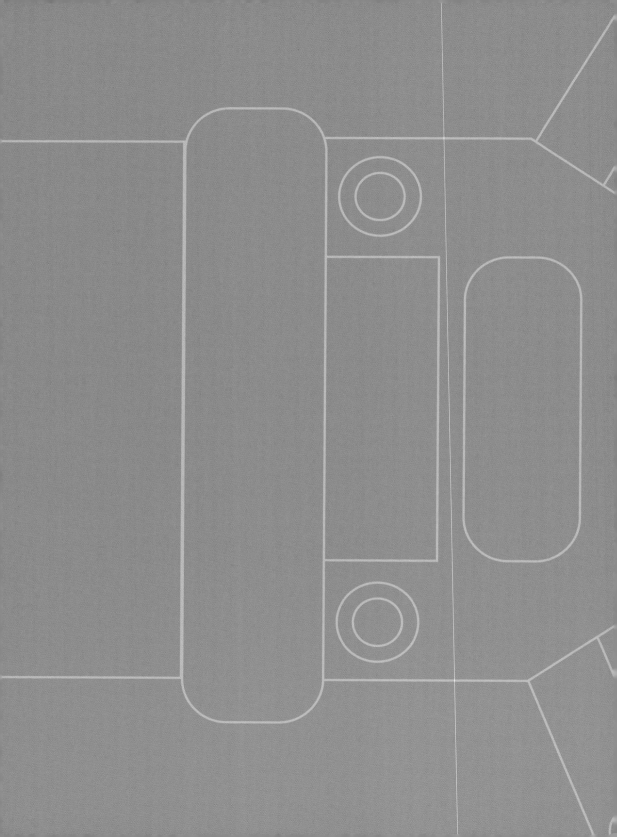

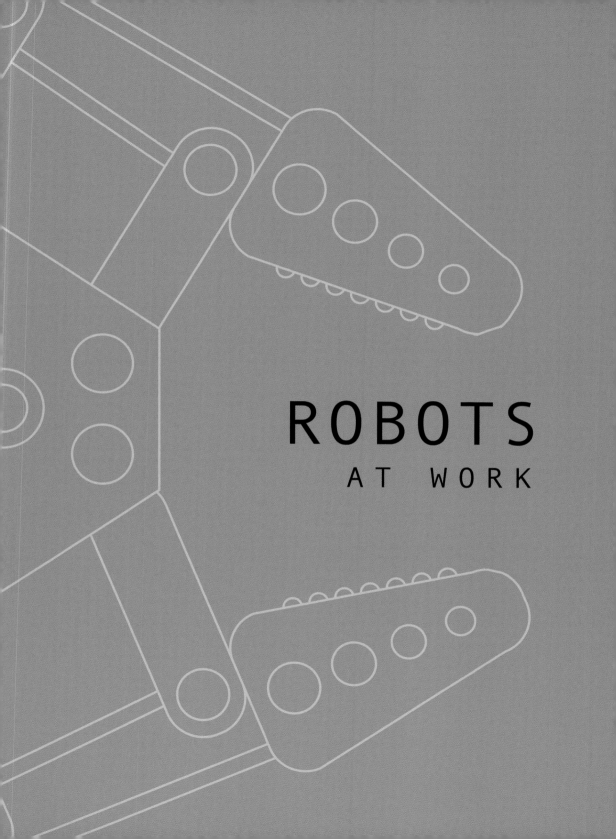

ROBOTS
AT WORK

ROBOTS AT WORK

Up until now, robots have tended to stay hidden away in factories, where they are the most efficient and reliable of workers. They have always been strong and fast, but they have also been half-blind and unable to make their way around on their own. Such factors limited their uses to a narrow range of tasks, but the situation is changing, and changing fast.

The first industrial robots were dangerous to be around and so were segregated from human workers. New industrial production-line machines such as FANUC's sensor-laden CR-35iA can work safely alongside people to combine the benefits of machine strength with human skill. A similar, but much smaller, robotic arm, Universal Robots' UR-10 takes the idea even further, with its makers daring to boast that its machines can automate any task.

Agriculture has been largely mechanised for decades, but there is still one 'last mile' to go before we can leave all the work to the machines. Cattle are rounded up by all-terrain vehicles (ATVs) rather than by horse, but the vehicle still needs a 'rider'. Cows are milked by machine, but the machines need an operator. And some crops, such as fruit, still need to be picked by hand. Ingenious inventors are working on all of these: the Lely Astronaut is a robotic cow-milking system, SwagBot is a robot cowboy, while the Agrobot picks the softest and most delicate fruit quickly and carefully using robotic arms.

Though robots are good at dull and routine tasks, there are several jobs that, though unskilled, always seemed too tricky for a robot to master. But not anymore. Now a GEKKO Facade Robot can clean skyscraper windows and the Alpha burger-bot prepares gourmet hamburgers, assembling and bagging them ready to go. If you bought this book online, then it was probably handled by an Amazon Kiva robot somewhere along the way. Meanwhile, in the skies above us, our planes are getting a makeover: the PIBOT is preparing to take over from flesh-and-blood pilots – if the passengers will accept it.

Not all jobs require hands or manipulators: many are just a question of inspection, and robots can get to places that are inaccessible to humans. The PureRobotics™ Pipeline Inspection System can travel down the insides of pipes and check their condition remotely, like a suburban version of a

Mars rover. Drones enable us to see landscapes in realtime for surveys, and Airobotics' Optimus 'drone-in-a-box' provides an automated system that can carry out detailed 3D surveys without human intervention.

In Karel Capek's 1920 play, R.U.R., the robots – inevitably perhaps – revolted against their human masters. This followed a deep-seated archetype, chiming with earlier fictional creations like Mary Shelley's Frankenstein. While there is no risk of the robots described here rising up, working robots might still trigger a revolution as they take over roles traditionally performed by us in everyday life.

Machines that take jobs from humans have always been contentious. In the eighteenth century, machine-breakers and Luddites tried to prevent the mechanisation of weaving by destroying the spinning mills. This view now looks faulty; mechanisation created as many jobs as it took, and even after two hundred years of automation there are still plenty of jobs for humans. Lamplighters, office messengers and broom makers may no longer exist, but their descendants are website developers, digital-content managers and life coaches. According to one estimate, as many as one-third of the children now in school will end up doing jobs that do not even exist yet.

Despite fears of robot workforces taking over, it is important to remember that labour is not as romantic as we might like to think but is often gruelling and dangerous. Robots offer a future in which nobody needs to do these tasks. A person might choose to do jobs that could be done by machine – cooking their own hamburgers or picking their own fruit – because they find it enjoyable, not because they are forced to earn a living.

ROBOT CAN MOUNT A
VARIETY OF TOOLS

SOFT RUBBER SKIN,
GREEN COLOUR SIGNIFYING
IT IS SAFE FOR HUMANS
TO APPROACH

2.8M (9.2FT)

CR-35iA

Height	2.8m (9.2ft)
Weight	990kg (2,180lb)
Year	2016
Construction material	Steel
Main processor	Proprietary processors
Power source	External mains electricity

Industrial robots can be dangerous workmates. In January 1979, Robert Williams was working at a Ford factory in Flat Rock, Michigan, alongside robots moving car parts from one area of the factory to another. A 1mt (1t) robot with a mechanical arm had come to a halt because it had not found the parts it expected on a storage rack. Such pauses sometimes meant that the robot was confused, so Williams climbed on to the rack to collect a part himself. The robot suddenly started up again, and its arm struck Williams in the head, killing him instantly. He was the first man reported to have been killed by a robot. A judge awarded $10m in compensation from the robot's makers because of the lack of safety measures.

Today, heightened safety awareness means that industrial robots are fenced off from their human coworkers behind barriers and warning signs. If a production line needs human intervention, the machines are shut down whenever a person is within reach. This forced separation makes it difficult for people and robots to work together. In recent years, Japanese company FANUC has developed a new type of industrial robot to collaborate safely with humans. With over four hundred thousand machines installed worldwide, FANUC dominates the field of industrial robotics, so is well placed to rise to this challenge. The showpiece of its collaborative robot range, CR-35iA, is a full-size robotic arm almost 2m (6.5ft) tall and weighing 990kg (2,180lb), and with six flexible joints that can lift 35kg (77lb).

The most obvious difference from other FANUC robots is that this one is green – signifying that it is safe to approach – rather than the company's traditional warning yellow.

Earlier robots were dangerous because their speed, strength and lack of sensors meant they could strike and injure humans without even noticing. This will no longer be the case with CR-35iA. Firstly, it is equipped with a highly sensitive safety sensor. This detects even the slightest contact between any part of the robot and all else in any direction. A 'contact stop' means it freezes instantly when it touches anything unexpectedly. Additional dual channel safety (DCS) software can monitor the distance to surfaces and ensure that the robot stops whenever it gets too close to them, leaving large enough gaps that people cannot be trapped. Secondly, the CR-35iA is covered in soft rubber skin so there are no sharp edges. Thirdly, the CR-35iA responds to touch: if it gets too close, you can simply push it away.

The thinking behind collaborative robots is to create machines which can safely work side-by-side with humans. The CR-35iA can use a built-in vision system to detect and pick up a specific part from a bin and hold it up while a human bolts the part into place. The robot does the simple tasks and heavy lifting, while the human handles the parts that require dexterity or judgement. The result? Efficient teamwork that combines human brains with robot brawn.

Collaborative robots have the potential to inhabit factory spaces previously unknown to robots. In the future, while factories may still use yellow industrial robots in fenced-off zones, collaborative robots – or at least robots designed to be harmless to humans – are likely to become increasingly common. They will help with a wider range of tasks and could eventually be introduced to working environments in which strength is necessary but safety issues have prevented the use of robots. Building sites, warehouses and garages are obvious examples, but a collaborative robot might also be useful in a hospital where patients need to be moved around.

Robot-related deaths receive a disproportionate amount of media attention, yet the number of deaths caused annually by industrial robots is tiny – around twenty-five in the United States – which accounts for 0.01 per cent of all industrial casualties. The CR-35iA is just the first generation of collaborative robots. Subsequent generations will be smarter and with better sensors. So while the biggest problem now facing robots sharing industrial spaces with humans may be human perception rather than safety issues, perhaps collaborative robots will help to change that too.

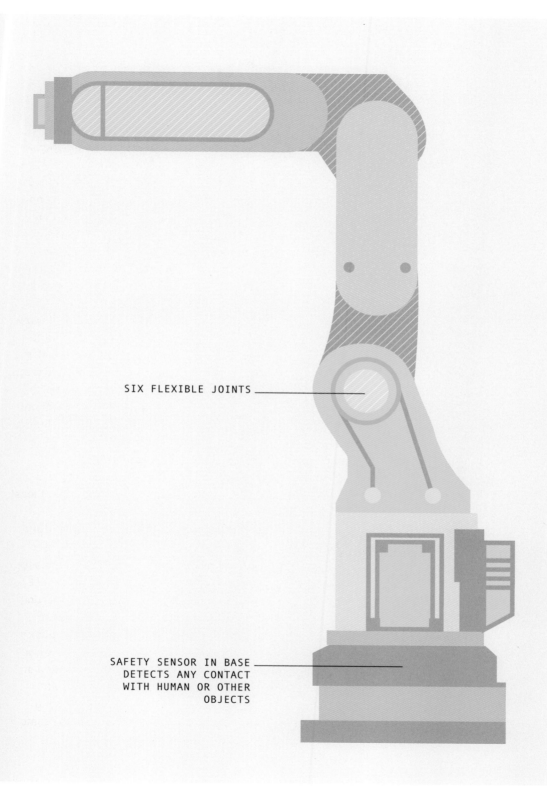

SIX FLEXIBLE JOINTS

SAFETY SENSOR IN BASE
DETECTS ANY CONTACT
WITH HUMAN OR OTHER
OBJECTS

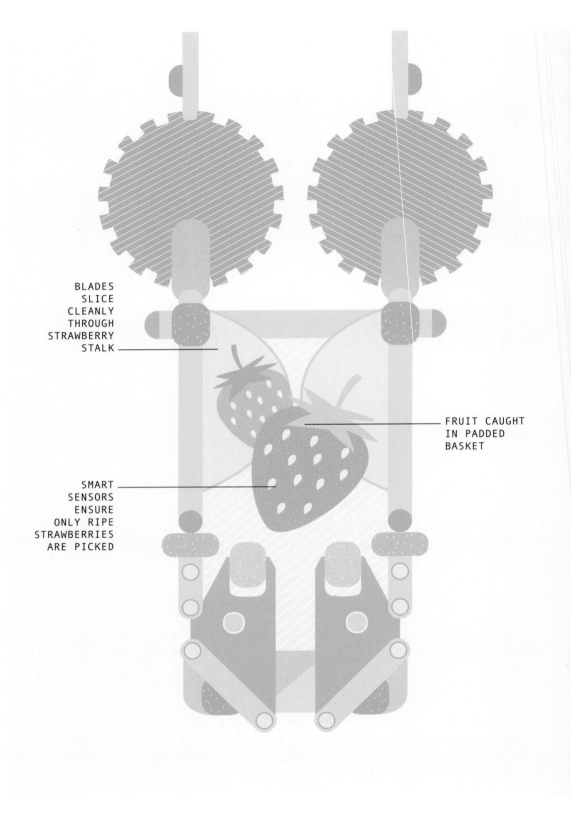

BLADES SLICE CLEANLY THROUGH STRAWBERRY STALK

FRUIT CAUGHT IN PADDED BASKET

SMART SENSORS ENSURE ONLY RIPE STRAWBERRIES ARE PICKED

AGROBOT SW 6010

Height	1.5m (4.9ft)
Weight	4mt (4.4t) approx.
Year	2015
Construction material	Steel, plastic
Main processor	Proprietary processors
Power source	Lombardini diesel 28.5HP engine

Fruit picking is one of the most labour-intensive jobs in agriculture. While a farmer can rapidly harvest an entire field of cereals with a combine harvester, fruit picking is still very much a manual occupation.

Traditionally, the fruit-picking season involves vast armies of migrant labourers coming into an area for the duration. In the United States, fruit pickers come from Mexico; in Europe they come from Eastern Europe or from North Africa. Movement of labour brings increasing challenges, and employers are starting to find it difficult to get enough hands at the low wages the work offers. Robots may provide an alternative.

Agrobot is a company based in the Spanish town of Huelva, the heart of the country's strawberry-growing region. Their Agrobot SW 6010 is the world's first fully automated strawberry-picker.

This might seem a simple, repetitive task, but gathering strawberries is not quite like working on an assembly line. It takes two particular skills that robots have trouble mastering: first, it has to identify the fruit among the foliage and determine whether a given strawberry is ripe enough to pick; second, it has to pick the fruit without harming it. This is especially difficult with strawberries, because there is no after-ripening: they have to be picked at exactly the right time, and any bruising causes them to rot.

The SW 6060 resembles a tractor. A diesel engine powers a hydraulic system that drives the wheels as well as the robot's picking arms. Each of the four large wheels has an independent hydraulic motor, and independent steering. This allows the robot to manoeuvre more easily in the cramped conditions of strawberry greenhouses, guided row by row by an autonomous navigation system. The external parts of the SW 6010 are plastic rather than metal. This is partly owing to weight and cost considerations, but also because plastic does not need grease or other lubricants that can attract dust and dirt.

The SW 6010 has a high clearance to drive over the fruit beds, while five robotic arms on each side reach down to harvest the fruit. Each arm is equipped with a set of sensors: a camera for inspecting the fruit, ultrasonic sensors to keep the arms a safe distance from the ground and from other arms, and inductive sensors tracking the position of each arm. Software ensures that the movement of the arms is synchronised with the robot's steady forward progress.

A proprietary, artificial intelligence (AI) vision system, AGvision, has been developed to assess the fruit. Taking twenty images per second, it works on size, shape and colour, analysing each fruit individually. Fortunately, nature helps with colour coding: ripe fruits are a vivid red, while unripe fruits are shades of pale green. If a strawberry is judged ripe enough, a pair of razor-sharp blades cuts through the strawberry's stem, and the fruit is caught in a padded basket. The arm then transfers the fruit to a conveyor belt on the body of the robot, which carries it into the packaging area.

Two operators riding the robot inspect the fruit and carry out the final packing in trays; the system is not yet quite as good as a human, so this final supervision is still needed. In the future the robot should be able to do everything by itself, delivering a finished product like a combine harvester.

Agrobot's success is likely to turn on economics. Fruit picking is tedious, backbreaking work, involving workers typically spending extended shifts stooped over a crop to complete the harvest before the end of the picking season. As labour becomes scarcer and more expensive, and as robots become increasingly efficient, the economic balance will tip in favour of the machines, allowing the Agrobot SW 6010 to step in as a labour-saving device that eliminates low-paid work. The only question is how long this will take to become a reality.

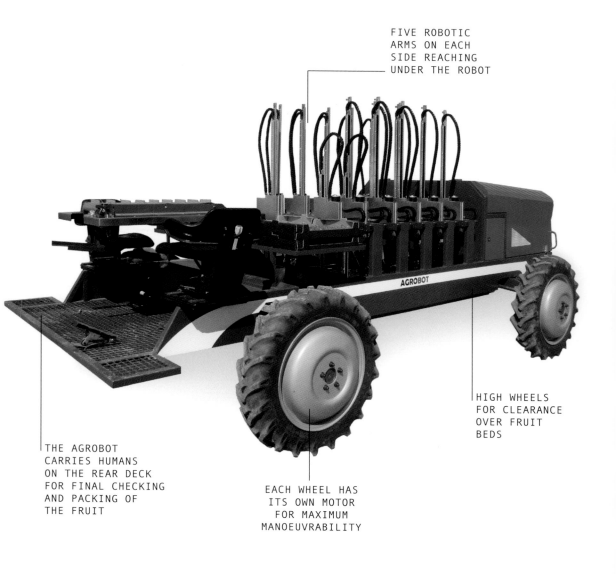

FIVE ROBOTIC
ARMS ON EACH
SIDE REACHING
UNDER THE ROBOT

HIGH WHEELS
FOR CLEARANCE
OVER FRUIT
BEDS

THE AGROBOT
CARRIES HUMANS
ON THE REAR DECK
FOR FINAL CHECKING
AND PACKING OF
THE FRUIT

EACH WHEEL HAS
ITS OWN MOTOR
FOR MAXIMUM
MANOEUVRABILITY

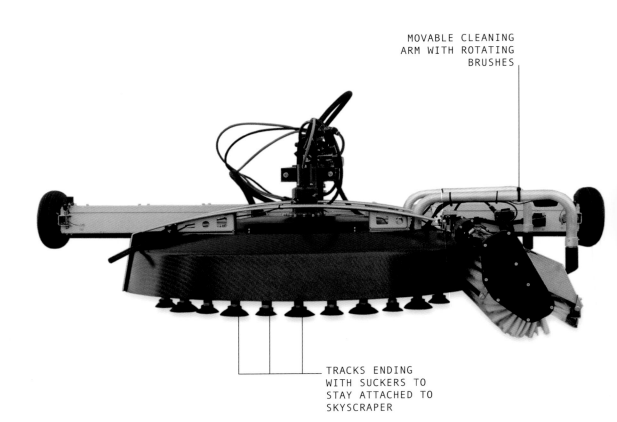

MOVABLE CLEANING
ARM WITH ROTATING
BRUSHES

TRACKS ENDING
WITH SUCKERS TO
STAY ATTACHED TO
SKYSCRAPER

1.38M (4.54FT)

GEKKO FACADE ROBOT

Height	42cm (16.5in)
Weight	70kg (154lb)
Year	2015
Construction material	Plastic
Main processor	Commercial processors
Power source	External mains electricity

City centres across the globe are taking on the same look: whether in the Far East, the heart of Europe or the United States, each now boasts a cluster of glass-sided skyscrapers as a badge of its commercial success. Window cleaning, however, has failed to keep pace with advances in architecture, and workers wielding sponge mops still work from suspended platforms to keep the vast areas of glass free of city grime – until now, that is, with the arrival of the GEKKO Facade Robot.

Invented by Swiss company Serbot AG, the GEKKO Facade Robot is the world's first window-cleaning robot for large vertical surfaces, provided as part of a package of building care. Like a human window cleaner, a GEKKO is lowered from the top of a building on ropes, and has a hose connection to a water supply and suction cups for stability. The difference lies in the speed and agility with which the GEKKO works. The disc-shaped robot has two tracks, with ten suckers each, that 'walk' along the face of the building, keeping the robot securely attached, even when buffeted by the wind. The GEKKO does not need guide rails or other aids to find its way around, and its traction allows it to climb over horizontal and vertical surfaces, inclines and even overhangs that are impossible to reach by normal methods.

The GEKKO can raise or lower its cleaning arm to bring it into contact with the window glass. The arm contains a series of rotating brushes, like those of a carwash or street cleaner, which do a more thorough job than manual cleaning. The makers claim that it uses less water than traditional cleaning methods, and the powerful brushes mean that no detergent is required – so it is ecofriendly, to boot.

The benefits of the GEKKO Facade Robot are manifold. Operators can control the robot using a joystick, or can set it to function automatically. This means that the manpower needs are minimal, and it is possible for a single worker to clean an entire skyscraper. With GEKKO cleaning 600m² (720sq yd) of glass in an hour, it is around fifteen times as quick as a typical human. And, of course, there is no need for tea breaks, but the potential for longer shifts. When the cost of cleaning a large building can run to eighty thousand pounds, these benefits can amount to significant savings.

Not only that, but the window cleaning of tall buildings is inherently dangerous. Even when there is no perceptible breeze at ground level, the wind may be gusting at 30mph (48kmh) one hundred storeys up. Skyscrapers towering 305m (1,000ft) experience powerful winds, and require giant internal pendulums, called tuned mass dampers, to prevent them from swaying perceptibly. The occupants may never be aware of the windspeed outside, but window cleaners feel the full force and accidents are common. Thanks to its suction feet, the GEKKO can operate in winds that would keep human cleaners grounded.

There is also a growing awareness of security in the business world. Passwords and firewalls may be mandatory, but if a window cleaner appears outside a meeting room during a presentation of plans for a new product launch, then such precautions may be wasted. GEKKO may ease such concerns. The robot also guarantees discretion when cleaning high-rise apartment buildings or hotels where guests may leave the curtains open to enjoy the view. This may be one of the few areas in which we are more comfortable with machines than humans.

There is no sign of the trend for glass-walled skyscrapers abating, and the tallest buildings still attract a premium. Architects are aware of the need to create landmark buildings with unusual shapes, which create challenges for traditional cleaning methods. The GEKKO also scores here, specifically on the quality of its work. By their nature, skyscrapers tend to be prestige properties, and companies pay a hefty premium for showpiece offices with spectacular views. Any dirt stuck in the window rims undermines the whole effect. GEKKO's rotating brushes and machine-consistent cleaning ensures that not one corner is missed. The days of hearing someone say 'you missed a bit' could be over as the GEKKO finds its habitat expanding in the coming decades.

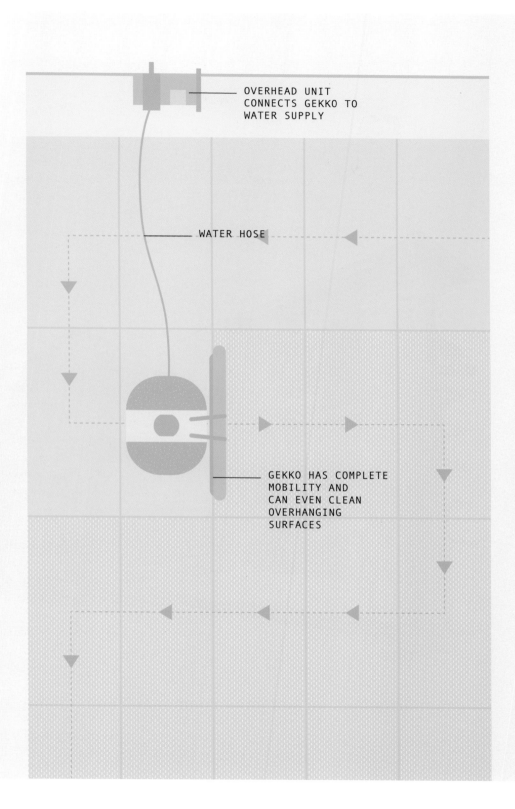

OVERHEAD UNIT
CONNECTS GEKKO TO
WATER SUPPLY

WATER HOSE

GEKKO HAS COMPLETE
MOBILITY AND
CAN EVEN CLEAN
OVERHANGING
SURFACES

LASER SCANNER
READS TAG ON COW'S
EAR TO IDENTIFY

2.37M (7.75FT)

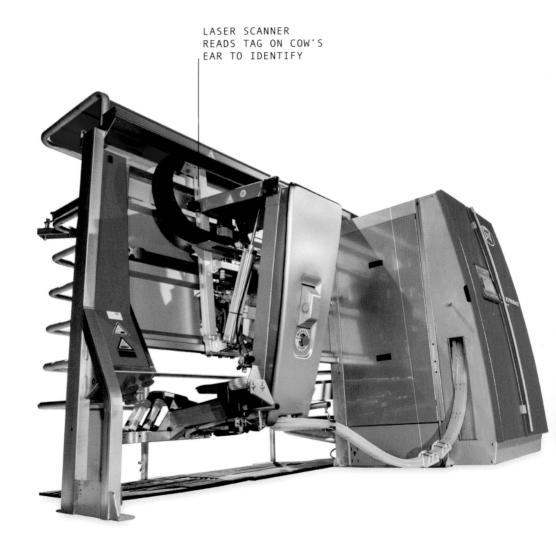

LELY ASTRONAUT A4

Height	2.37m (7.75ft)
Weight	650kg (1,430lb)
Year	2010
Construction material	Steel
Main processor	Proprietary processors
Power source	External mains electricity

Drinking milk is in our DNA: thousands of years ago, around the time we domesticated cattle, humans in Europe and North Africa acquired a mutation giving adults tolerance to lactose. Ever since then, people have spent long hours milking by hand. Milking machines were invented in the mid-nineteenth century, but it took decades to work out a practical way of drawing milk from the cow using suction cups.

Now milking machines are universal, and the labour requirement has dropped dramatically, but the twice-a-day routine is still one of the most loathed jobs on the farm. A farmer has to rise before dawn, round up the herd, get it to the milking shed, and then go through the process of attaching each cow to the milking machine, milking and then detaching again, one by one. The whole palaver repeats just twelve hours later.

The Astronaut A4, made by Dutch company Lely, automates the entire process, even getting the cows to come in to the milking shed in the first place. And when it comes to milking, the Astronaut represents the state of the art on applying robotics to the task. As each cow enters the shed, an RF radar scans the tag around her neck – just like a supermarket barcode scanner – and she is offered a trough of feed. If the ID on the tag shows she has been milked recently, the trough is withdrawn and the cow can leave the unit. Otherwise the cow is given an amount of feed tailored to her requirements. This treat motivates the cows to come in for milking.

Once a cow is in place at her feeding trough, the robotics come into play. A 3D depth camera tracks the cow's movement and a robot arm, guided by a laser scanner (a 'three-level teat detection system'), cleans the teats and attaches suction cups. After milking, the teats are treated, and the robot arm places the suction cups in a steam cleaner to prepare them for the next cow.

Cows are herd animals and prefer to stay together. Unlike other milking machines, the Astronaut does not need the cows to be in individual stalls but has an open-plan arrangement that keeps them within sight of each other. The 3D camera follows a cow so that, if she moves, the robot arm and the suction cups simply move with her. The cows feel safer and more comfortable, and the milking process goes more easily.

As with other milking machines, the Astronaut records the volume of liquid with each milking and maintains a database of the cow's production history, so that diet and medication can be adjusted accordingly. Astronaut also analyses the milk as it is collected, measuring the fat content, protein and lactose and checking for signs of illness. The Astronaut alerts the farmer if any of the cows in the herd has missed a milking or when the cow requires specifal care.

Lely's slogan for Astronaut is 'the natural way to milk'. The whole automated, computerised process might seem dauntingly technological, but it is carefully crafted to fit in with the cow's needs. Unlike the traditional arrangement, the Astronaut puts the cow in charge of the milking process, and gains the benefit of the animal's cooperation.

A key consideration for robot milking machines is the cost. Farm jobs tend to be carried out manually by low-paid labourers, often immigrants, until a tipping point is reached, at which it is cheaper for a machine to do the job. An Astronaut can cost around £100,000, and can manage a herd of 70 cows. It is not a cheap option, but as old milking machines are being replaced anyway, the additional cost might not be so significant.

Lely describes the Astronaut as 'the most reliable employee imaginable' that never misses a day's work and patiently gives every cow individual attention from first to last. It may not fit with our traditional idea of dairy farming, but the cows like it. As do farmers: adoption is catching up rapidly. In some countries, up to 50 per cent of new investments in milking are in robotics.

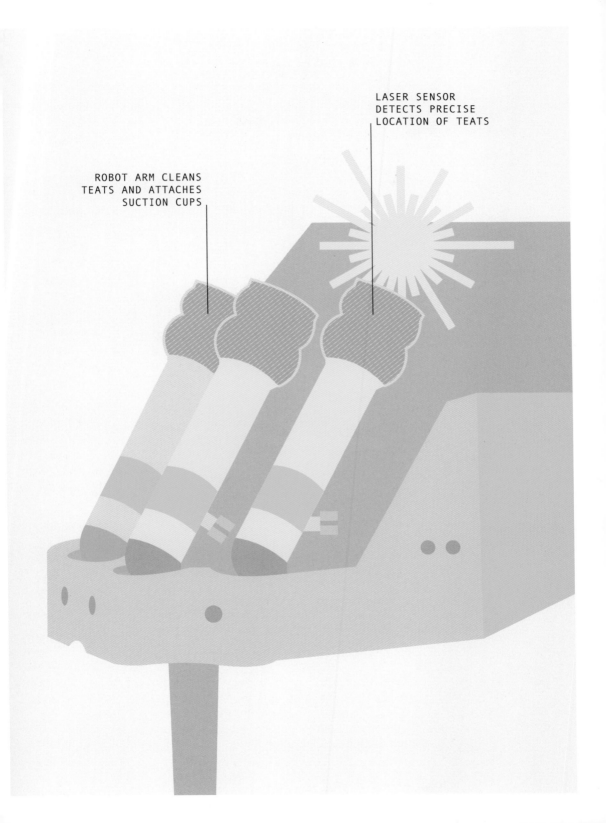

LASER SENSOR
DETECTS PRECISE
LOCATION OF TEATS

ROBOT ARM CLEANS
TEATS AND ATTACHES
SUCTION CUPS

KIVA'S LIFTING MECHANISM
THIS DEVICE IS PLACED UNDERNEATH
EACH STORAGE POD AND THEN ROTATED TO
LIFT THE STRUCTURE UP

CORKSCREW ROTATION
ALLOWS KIVA TO PICK
UP AND CARRY

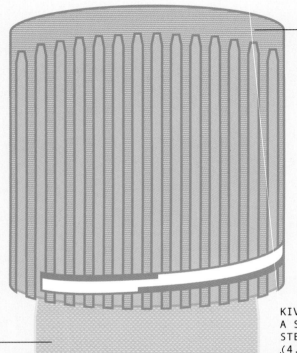

LIFTING SYSTEM
ALLOWS IT TO
CARRY WEIGHTS
UP TO 4X ITS
OWN WEIGHT

POD IS COUNTER-
ROTATED TO AVOID
IMBALANCE

KIVAS MOVE AT
A SLOW AND
STEADY 3MPH
(4.8KMH)

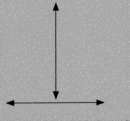

MOVE IN STRAIGHT LINES

TURN IN PERFECT RIGHT ANGLES

KIVA

Height	30cm (12in)
Weight	110kg (243lb)
Year	2005
Construction material	Steel
Main processor	Commercial processors
Power source	Lead-acid battery

Bright-orange Kiva robots are the worker ants of online commerce giant Amazon's warehouse ecosystem. Constantly in motion and seemingly moving at random and always hauling heavy loads around, they ensure that millions of Amazon orders go out every day.

These orange bots descend from prototypes built by Kiva Systems of Woburn, Massachusetts, in 2005. Each robot has a base about 60 x 80cm (24 x 31in) and 30cm (12in) high. On top is a lifting mechanism that works like a corkscrew. It allows the Kiva to slide underneath a 'pod' (a shelving unit loaded with goods), jack it up and cart it away like a porter carrying a load on on his or her head. Pods are 1m² (11sq ft) and 2m (6.5ft) high and can weigh 400kg (880lb) – four times as much as the Kiva. Travelling at 3mph (4.8kmh), the robots move carefully in straight lines, making right-angled turns. With each turn, the lifting mechanism counter-rotates so that the pod does not move. This reduces the chance of the pod overbalancing or items slipping off.

The process of fulfilling an order is known as 'picking, packing and shipping'. Before the robots came, human packers had to walk endlessly up and down aisles of shelves, finding and picking each item and bringing it back to a packing station. Now the humans remain at their stations, and the Kivas bring the shelving pods with the items needed for each order to them.

The Kivas take instruction from a central routing system. Each robot keeps track of its position by scanning barcode stickers placed on the floor at 2m (6.5ft) intervals. It broadcasts its own position via Wi-Fi. Scanners also confirm the identity of a pod before the robot picks it up.

When a robot lines its pod up in front of a packer, the exact item to be selected is highlighted with a laser, making it easier for the packer to get the right one. Once the order has been fulfilled, the Kiva returns the pod back to its place in the warehouse.

In addition to laser scanners, each Kiva has obstacle avoidance sensors. In theory, these should not be needed: humans should never be in the highways roamed by Kivas, and because every robot's location is known, they should never collide with one another. But items do occasionally fall from pods and block the way, and collision avoidance is essential if humans ever do stray.

Kiva robots run on lead acid batteries, taking breaks every two hours to recharge themselves automatically. While lithium ion batteries would give a longer running time, the cheaper lead acid batteries are favoured because a recharger is never far away.

Amazon acquired Kiva Robotics in 2012, and since that time the company has supplied them exclusively. In 2012, Amazon had around five thousand robots; by 2017 they had more than forty-five thousand. The takeover shook up the warehousing robot market, as previous Kiva customers, including office supply companies Staples and Office Depot, and clothing company Gap, had to find new suppliers.

Like worker ants, Kiva robots are tireless, hardworking, uncomplaining and do not need to be paid. According to studies, the robots increase the rate of picking, packing and shipping orders at an Amazon warehouse by a factor of between two and six, while also decreasing the number of errors. As Amazon and other businesses operating on a similar model expand, so too will the armies of Kiva-style robots.

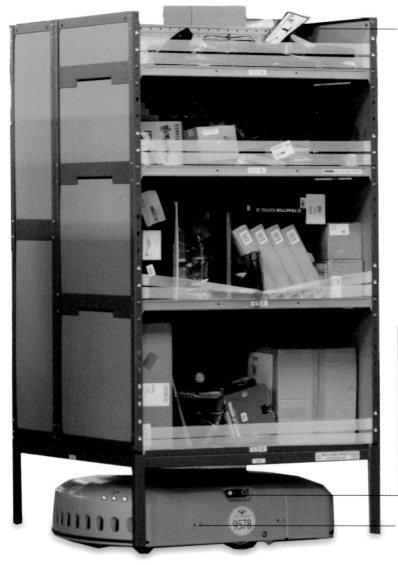

EACH POD IS DROPPED OFF IN A QUEUE IN FRONT OF ITS PACKER BEFORE BEING RETURNED TO STORAGE

KIVAS USE A SCANNING DEVICE TO ENSURE THAT THEY COLLECT AND DELIVER THE CORRECT PODS

KIVAS ARE ABLE TO AVOID ACCIDENTS OR MISHAPS DUE TO THEIR OBSTACLE AVOIDANCE SCANNERS

30CM (12IN)

80CM (2.6FT)

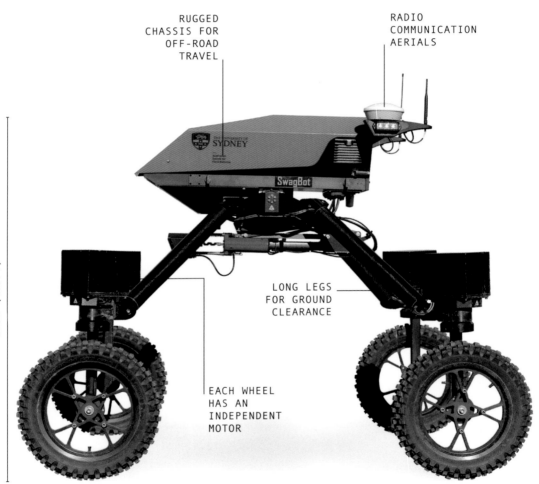

RUGGED
CHASSIS FOR
OFF-ROAD
TRAVEL

RADIO
COMMUNICATION
AERIALS

1.5M (4.9FT)

LONG LEGS
FOR GROUND
CLEARANCE

EACH WHEEL
HAS AN
INDEPENDENT
MOTOR

SWAGBOT

Height	1.5m (4.9ft)
Weight	190kg (420lb)
Year	2016
Construction material	Steel
Main processor	Proprietary processors
Power source	lithium ion battery

In Australia, the average cattle station – they are never called ranches – spreads out over 1,550sq mi (4,000sq km), making it three times the size of Greater London. The arid land cannot support many cattle, with perhaps one animal per several hectares compared to Europe's several animals per hectare (2.5 acres). With their cattle spread out over such immense distances, cattle stations need a supply of cowhands to help with rounding up – they need SwagBot.

The stereotypical farm worker is a swagman, so-called because he carries all his belongings in a roll or swag, offering his services in exchange for food and board. Swagmen are now in short supply, so Salah Sukkarieh of the University of Sydney is leading a team to develop a robot to take on farmwork, including rounding up cattle. Rather than trying to turn a jeep or quad bike into an autonomous vehicle, they started with a clean slate: 'We built a new platform rather than adapt an ATV or quad bike because the aim is to develop fully autonomous capability with high-terrain adaptability and manoeuvrability', explains Sukkarieh.

With four long legs on wheels – each with its own motor – SwagBot is designed for high mobility. It is omnidirectional (able to travel in any direction), and has a rugged composite chassis to absorb any battering from cross-country travel. The drive system is waterproof, so SwagBot can travel through water; in principle, it could even submerge completely.

SwagBot can make its way across typical cattle station terrain, negotiating ditches, streams, logs and other obstacles, with a top speed of 12.5mph (20kmh). SwagBot can also function as a robot tractor, pulling a trailer. The most important thing is that it does not get stuck and require rescuing, because it will be working so far from help.

In its first field trials, SwagBot successfully herded cattle by remote control. Having developed SwagBot as a mobility platform, Sukkarieh's team is now focusing on its software and sensors. The aim is for the robot to be able to traverse areas that have been mapped without human assistance. Sensors under development include a basic collision-avoidance device – traffic is not a big problem in the Outback – and a more sophisticated system that uses a video camera to assess the quality of the grazing and ensure there is enough food for cattle. Other sensors will examine the cattle themselves, judging their health with thermal imaging and using a camera to assess their gait for signs of lameness. SwagBot may also get a sensor that can sample dung to check on an animal's health and wellbeing.

More sensors and smarter software will extend SwagBot's abilities even further, perhaps to include autonomous delivery – to take a trailer of feed to a given location and leave it there, or retrieve a trailer, for example. Sukkarieh is also looking at using SwagBot for autonomous weeding. Spotting and destroying patches of invading vegetation prevents them from spreading. Swagbot may also act in coordination with other robots. One trial involved a small flying drone to directing SwagBot, which would act as a mobile base station and recharging unit for the drone. The drone could then help find missing cattle, or scout out the easiest route through difficult terrain ahead.

As usual, price is a key consideration here. Sukkarieh says that the falling cost of electronic components now makes something like SwagBot start to look commercially viable. It may not happen immediately, but at some point, cheaper robots with improved capability will start to cross over with the rising cost and limited availability of labour. This can only mean that, ultimately, the outback will belong to Swagbot.

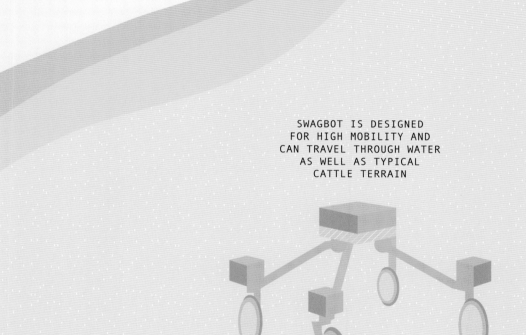

SWAGBOT IS DESIGNED
FOR HIGH MOBILITY AND
CAN TRAVEL THROUGH WATER
AS WELL AS TYPICAL
CATTLE TERRAIN

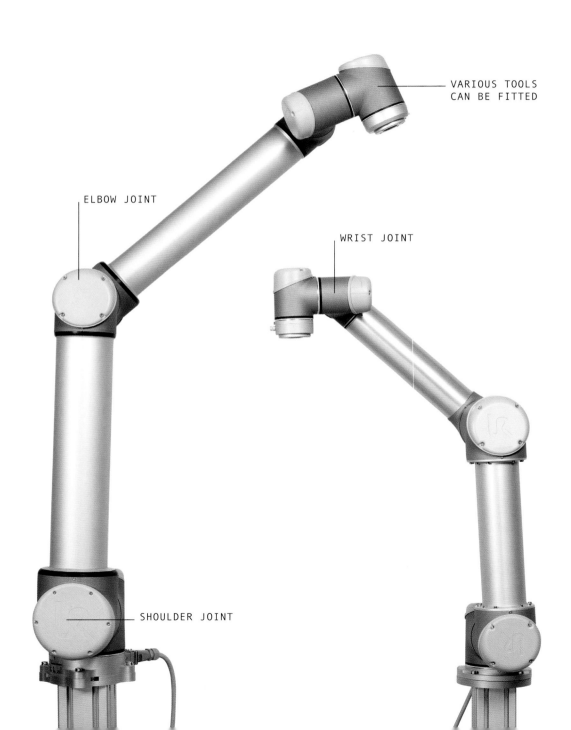

VARIOUS TOOLS
CAN BE FITTED

ELBOW JOINT

WRIST JOINT

SHOULDER JOINT

UNIVERSAL ROBOTS UR10

Height	1.4m (4.6ft)
Weight	28.9kg (64lb)
Year	2013
Construction material	Steel
Main processor	Commercial processors
Power source	External mains electricity

Industrial robots are generally big, expensive and complicated machines. The process of automating production can take months, often completely rebuilding a production line around the robot. This type of automation has worked well in some industries, but it is not suitable for the majority of businesses. Danish company Universal Robots (UR) – a name that recalls Karel Capek's original, fictional Rossum's Universal Robots – wants to change that. Its range of small robots applies automation to practically every human activity: 'When we say the Universal Robot can automate virtually anything, we mean virtually anything,' claims the company website.

The UR machines are smaller than traditional industrial robots. The biggest is the UR10, which weighs just 28.9kg (64lb). Their diminutive size makes them safer to be around than their larger cousins, which have the potential to kill a person with a single wrong movement. Most UR robots work in areas shared with humans without the need for safety barriers.

Physically, the UR10 looks like other industrial robots. It is stationary and resembles an Anglepoise lamp with six 'degrees of freedom': the shoulder, elbow and wrist joints rotate six different ways so that the robot can move around as needed. A UR robot can also hold a wide variety of tools and devices.

What sets the UR robots apart from others is their user interface. Rather than requiring a dedicated team of engineers, UR robots can be set up by an operator with no programming experience. Programming is carried out on a tablet computer with a touchscreen, and by moving the robot arm to the positions it needs to take up during the task. Effectively, the operator shows the robot what it needs to do. UR claims the average set-up time for a customer is half a day. An untrained operator can open the box, get the machine set up, and program it for a simple task in less than an hour.

Its small size and light weight makes the UR10 suitable for light factory work rather than welding cargo ships, and gives it tremendous flexibility. When a robot is needed for a new task, an operator can literally pick up a UR10, carry it to another part of the factory and get it working without any assistance.

UR robots perform the same kinds of tasks as traditional industrial robots. At Renault's plant in Cleon in France, UR10s drive screws into engines; the robot's flexibility and size allows it to reach places that are hard for human workers to access. After securing each screw, the UR10 checks and verifies it with robotic thoroughness.

Other companies use UR machines for 'pick and place' jobs, selecting and positioning parts with the guidance of a vision system. Robots are also used for assembly and packing. At Xiamen Runner, one of the world's largest manufacturers of bathroom fittings, UR10 robots operate injection-moulding machines to produce components. The company mounts its robots on rails to move them from one job to the next. The quick setup means the UR10s are suited to producing customised products in low volumes, rather than in large runs. This is a shift away from the system of mass production that has been in place since the days of Henry Ford.

UR claims its robots have the fastest payback time of any in the industry, paying for themselves in a little over six months. The ease with which UR machines can be applied to existing operations, without safety issues or the complexities of programming, suggests that this type of machine could spread far and wide. And don't be fooled into thinking that UR machines are limited to industrial environments. Every secondary school in Denmark has a small UR3 robot for teaching technology studies, and Mofongo's Distillery in Groningen in the Netherlands has a UR robot bartender that slides up and down a rail pouring and serving drinks.

Apple, now the world's most successful company, built its success by making products with highly intuitive user interfaces. The UR machine may turn out to be the robot equivalent of the iPhone, a mass-market machine with universal appeal.

THE UR10 IS EASILY
PROGRAMMED VIA A
TABLET CONTROLLER AND
MOVING THE ARM TO THE
POSITIONS NEEDED

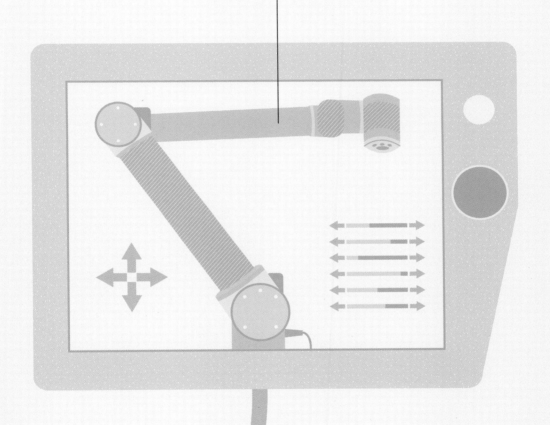

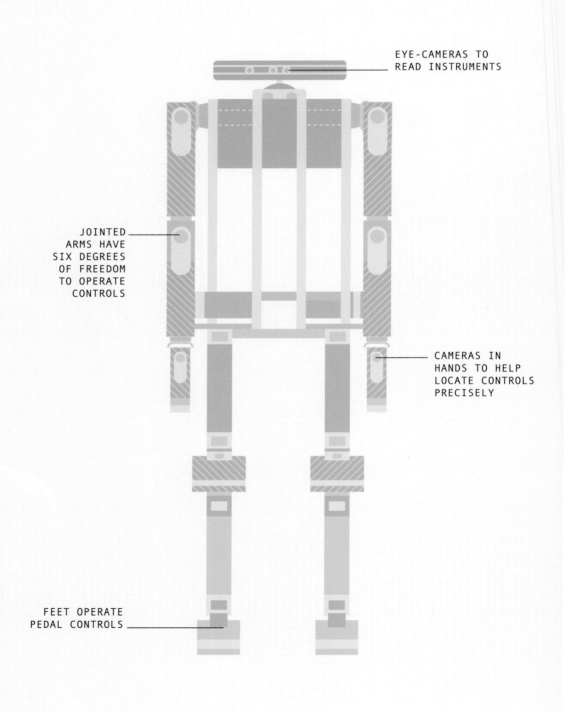

EYE-CAMERAS TO
READ INSTRUMENTS

JOINTED
ARMS HAVE
SIX DEGREES
OF FREEDOM
TO OPERATE
CONTROLS

CAMERAS IN
HANDS TO HELP
LOCATE CONTROLS
PRECISELY

FEET OPERATE
PEDAL CONTROLS

PIBOT

Height	1.27m (4.2ft), seated
Weight	24kg (53lb)
Year	2016
Construction material	Steel
Main processor	Intel NUC5 17 RYH
Power source	Battery

There are no prizes for guessing that PIBOT, short for PIlot roBOT, is a humanoid robot that sits in the pilot's seat and flies an aeroplane. What you might not guess, however, is that this robot is not set on autopilot, but actually uses the controls, just as a human would.

To build PIBOT, Professor David Hyunchul Shim and his colleagues at KAIST (formerly Korea Advanced Institute of Science and Technology) took a systematic approach to automating the task of piloting. His team broke piloting down into three areas – recognition, decision and action. They then developed the necessary hardware, machine intelligence and sensory software for a robot to carry out each of these.

PIBOT operates the controls, including the throttle, stick and pedals, and reads the dials and gauges just as a human pilot would. Unlike most drones, PIBOT is not remote-controlled, but flies the plane itself without any human involvement. It even uses the radio to talk to air-traffic controllers giving the same information and responses as a human pilot.

First demonstrated in 2014, the original PIBOT (PIBOT 1) was a scaled-down version based on a low-cost commercial robot, the Bioloid Premium. It successfully flew a complete flight in a flight simulator, from turning on the engine and releasing the brakes, to taxiing, takeoff, flying a predetermined route, and finally landing safely at the destination. This small android also flew a model aircraft. Some human assistance was necessary for landing, as the vision software still needed tweaking, but the exercise demonstrated that the concept was sound.

PIBOT 2 is a full-size humanoid robot. It costs around $100,000 to build, and is a fully functioning replica of a pilot. The arms and legs have six degrees of freedom and the hands another five. As well as cameras for eyes, PIBOT has cameras in its hands to aid with locating the controls. Like the earlier version, it has proven itself on a flight simulator. Shim's team are now working through PIBOT's responses to emergency situations, especially those that are not preprogrammed.

There is already a demand for this type of machine. The US Air Force is looking for a 'drop-in robotic system' that users can install quickly without modifying an aircraft to convert if from manned to unmanned operation. Shim's technology provides the starting point. The Air Force plan to use the robots for routine cargo transport flights and refuelling missions. In the longer term, the robots will take on more challenging intelligence, surveillance and reconnaissance operations (ISR). Given that unmanned aircraft are likely to be sharing airspace with manned planes in the near future, rules and standards allowing machines to fly may need to change.

Pilot robots may also find a role in the commercial world. Current regulations require a pilot and a copilot for every flight – the latter to step in in an emergency and to provide a second opinion. Commercial planes used to have a third crew member, the flight engineer, who monitored the instruments and calculated fuel consumption, among other tasks. Automated systems have already replaced flight engineers, and copilots look set to go the same way. Pilots are skilled and highly paid individuals, making them expensive to hire and to train. PIBOT offers an alternative, with low costs and steady improvement with each software upgrade. The skill needed to fly a new type of aircraft is just a download away, and PIBOTs can move from one type to another without any risk of getting confused.

While it is unlikely that airliners will fly without an onboard human pilot any time soon, it is worth remembering that most air crashes are caused by human error. The commonest accident type, 'controlled flight into terrain', where a pilot runs into a mountain or hillside, usually happens because the pilot ignores or misunderstands instrument readings. In years to come, passengers may well feel safer knowing that there is a PIBOT in the cockpit rather than a fallible human.

PLUG AND PLAY: PIBOT OCCUPIES
THE PILOT'S SEAT AND CAN FLY
WITHOUT ANY MODIFICATIONS
NEEDED TO THE AIRCRAFT ITSELF

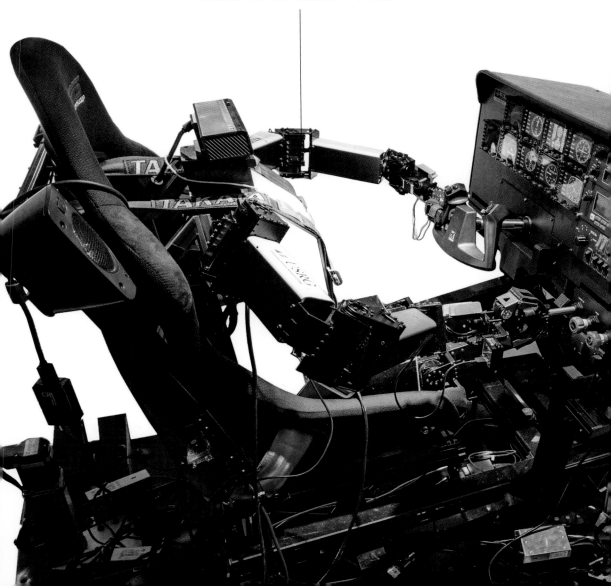

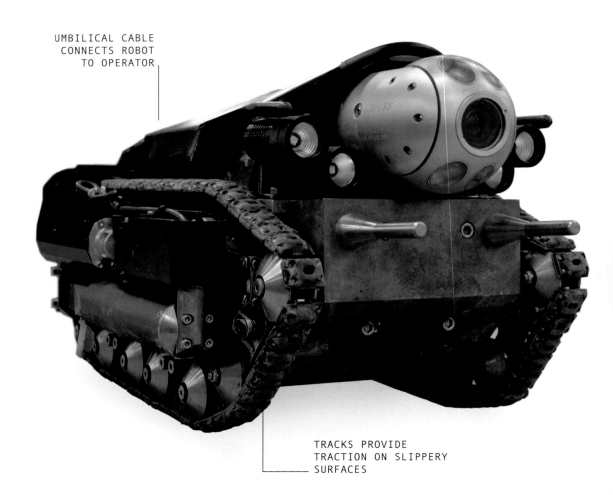

UMBILICAL CABLE
CONNECTS ROBOT
TO OPERATOR

TRACKS PROVIDE
TRACTION ON SLIPPERY
SURFACES

ROBOTIC PIPELINE INSPECTION SYSTEM

Height	25cm (9in)
Weight	135kg (298lb)
Year	2013
Construction material	Steel
Main processor	Commercial processors
Power source	Battery

Modern life is built on a mass of hidden pipelines that supply our utilities. They bring clean water into our homes, and take wastewater and sewage away. They silently transport gas and oil around the globe. Keeping the flow going and preventing leaks is essential.

Until the 1960s, pipelines could only be inspected from the outside. Preventive maintenance was impossible, except for pipes large enough for a human to enter, and problems could only be fixed when the pipe visibly cracked or sprung a leak. Then in 1965 'smart pigs' arrived to save the day.

Originally, these 'pigs' were bundles of straw, wrapped in wire. The pig matched the diameter of a given pipe and was sent through it to clean by scouring. The name 'pig' supposedly comes from the fact that they squeal as they go through pipes. There were also separator pigs, which acted as mobile plugs and allowed liquids such as fuel oil and crude oil to be sent down the same pipeline in succession. Someone then had the bright idea of putting cameras and other sensors on pigs to carry out inspections.

Able to detect corrosion and cracking, smart pigs are invaluable, using cameras or magnetic sensors able to detect corrosion and cracking. However, they also have their shortcomings. They are 'free swimming', moving with the flow, and this makes it difficult to get a clear view of trouble spots and potential issues. Not all pipelines are 'piggable'; many have sharp turns or changes in diameter that a pig cannot negotiate. These disadvantages have led to the development of a robot to reach places that even the smartest pigs cannot.

The PureRobotics™ Pipeline Inspection System looks like a miniature tank on two sturdy tracks. It is portable, and small enough to be lowered on a tripod through a standard 45cm (18in) manhole opening. Radio communication is impossible underground, so a Kevlar-reinforced fibre-optic umbilical cord carries data between the robot and the operator in a mobile control station on the surface from up to 2 miles (3.2km) away. The robot can work in pipes filled with water, but the makers recommend that pipes are 'dewatered' for the best results.

The front of the robot bristles with lights and high-definition cameras set in a rotating turret. It moves at around 1mph (1.6kmh), speed being less important than stopping for a good look. The main camera has a 10x zoom for close inspection. The robots records video, but also transmits it in real time to three screens in the mobile station, allowing operators to decide which areas require further investigation.

GPS and other navigation systems are useless underground where no radio waves penetrate, and there are no landmarks. Instead the robot has an inertial navigation system that calculates location from its velocity, acceleration and time travelled. This pinpoints a robot's position underground relative to where it started, and provides an accurate fix to guide diggers to where maintenance or repair is needed.

The maker, Pure Technologies Ltd, describe it as a system rather than a robot because it comes as a modular kit. For example, an additional tracked body attached to the basic unit doubles its size and carrying capacity in just a few minutes. The robot can tow a LIDAR sensor to scan a pipeline. LIDAR is an abbreviation of LIght Direction And Ranging, and is essentially a laser-based radar. A laser beam bounces off objects in the environment, and the time taken for the laser pulse to return indicates the distance. The LIDAR generates a 'point cloud' of dots around it; a computer can join these dots to build up a precise 3D map of an area. The robot can also carry magnetic sensors or sonar systems.

Exploring an inaccessible environment with scientific instruments, this Robotic Pipeline Inspection System carries out a similar task to NASA's Curiosity Rover. It may be less glamourous than space travel, but far more important when you need to turn on a tap or flush a toilet.

ROTATING TURRET
WITH HIGH-
RESOLUTION
CAMERAS

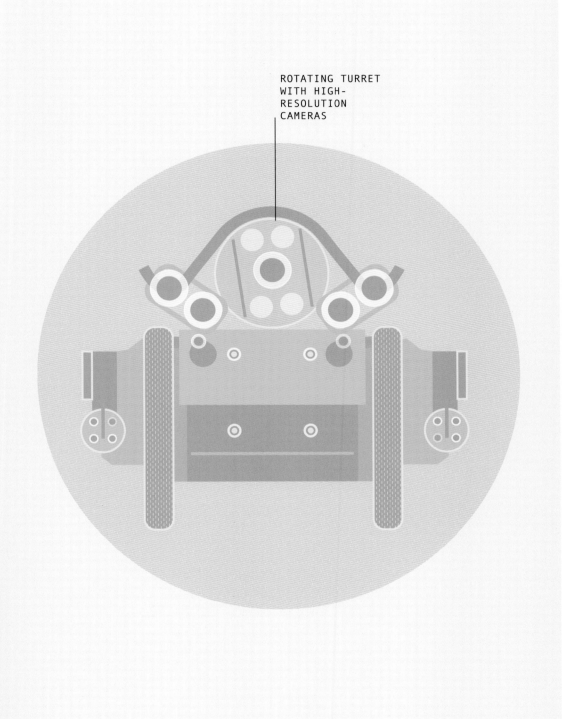

ROTARY OR
PNEUMATIC DRILL
DEPENDING ON
ROCK TYPE

PIT VIPER-275CA

Height	20.4m (70ft)
Weight	4mt (4.4t)
Year	2016
Construction material	Steel
Main processor	Commercial processors
Power source	Diesel engine

Over the decades, mechanisation has dramatically reduced the number of people working in mines. Gone are the days of hewing the rock face with a pick and shovel – today everything is done by giant machines. Visit a quarry, and the only humans you'll see are in tiny cabs atop monster vehicles. And now that robotics is making it possible to automate the machinery, even the cab has disappeared in new rigs, such as Epiroc's Pit Viper-275CA – the 'CA' stands for cabless automation.

Surface mining involves drilling and blasting. Rock is resistant to compression, but can be torn apart from the inside. The basic technique is to drill a hole, fill it with explosive, light the fuse and stand well back. These days, a computer calculates the number, size and pattern of the holes for optimal rock-breaking effect, and a rig carries out the drilling. One blast sends several thousand tonnes of rock cascading down into the pit, ready to be gathered up by excavators and trucks.

Swedish company Epiroc is a leader in mining machinery. Originally developed by parent company Atlas Copco before their split in 2018, their Pit Viper-275CA is a robotic blast-hole drilling rig, a tank-sized vehicle that runs on two broad caterpillar tracks. It travels at just over 1mph (1.6kmh), typical for such machines, and LIDAR sensors prevent it from running into obstacles, people or other vehicles. The Pit Viper can drill a 27cm (11in) diameter hole down to a depth of almost 60m (196ft). As the drill descends, an automated handling system adds new drill sections into the 'drill string' from a carousel.

Drill positioning is achieved using a precision hole-locating system. Traditional satellite navigation systems, such as GPS, are only accurate to within a few metres, so mining drills are guided by an enhanced version with additional signals from ground transmitters. This makes positioning accurate to within centimetres. The Pit Viper carries either a rotary drill or a pneumatic drill. The rotary drill relies on the force behind it – the Pit Viper can put some 34mt (37.5t) of its weight into the drill to crush rock. The pneumatic drill is a giant version of a DIY-ers hammer drill, with a hammer action powered by air pressure. Known as a down-the-hole drill, it has a miniature jackhammer that descends to the point of drilling. The drill shaft is a little wider than the drill bit, leaving space for the rock dust and fragments, known as cuttings, to be flushed out and blown to the surface using compressed air.

The BHP carried out its first test with a fully automated Epiroc's drill drill at a quarry near Perth in Western Australia in 2014. The machine moved to its start position, levelled itself and drilled its first hole in the designed spot. Then it extracted the drill string, disassembling it piece by piece, and moved on and completed a pattern of fifteen holes. This demonstration was significant because the rig was not programmed by a human operator but by a computerised mine automation system that is designed to manage the entire mining operation.

Epiroc is automating an increasing number of rigs at sites mining iron ore, coal, copper and other minerals. Operators now work in offices at operations centres that may be hundreds of miles away from the mines they oversee.

Robot drilling rigs like the Pit Viper are only part of the solution. There are also unmanned loaders, haulers and trucks, vehicles resembling giant yellow playground toys for handling the rock after the blast. All of them can be controlled remotely or by a mine automation system.

One of the big advantages of these machines is safety: keeping people off the site reduces the number of accidents. There are also big gains in efficiency and productivity. Automated drills are faster than a human operator, are more consistent, and the machinery is used for longer hours in the day. Satellite and LIDAR sensors means that night and day shifts are the same to these machines. Should developments continue along the same lines, it could be that future generations of miners never even visit the mines on which they work.

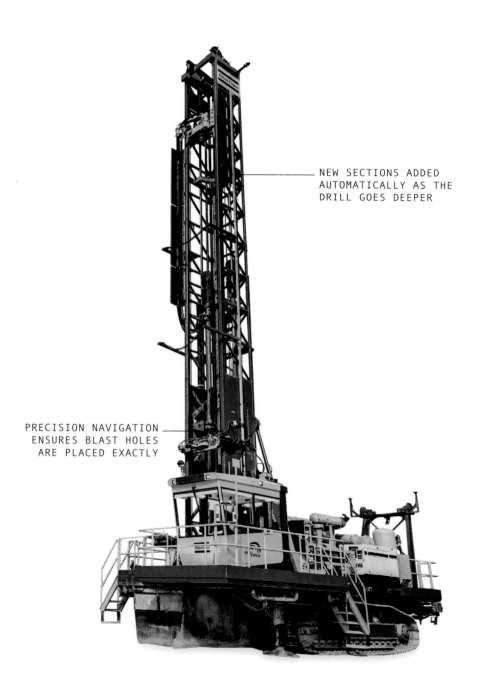

NEW SECTIONS ADDED
AUTOMATICALLY AS THE
DRILL GOES DEEPER

PRECISION NAVIGATION
ENSURES BLAST HOLES
ARE PLACED EXACTLY

20.4M (66.9FT)

SLIDING ROOF
PROVIDES
PROTECTION FROM
WEATHER

ROBOT ARM
INSIDE
AIRBASE
CHANGES
BATTERIES
AND SENSORS
AS NEEDED

AIRBASE WITH
COMMUNICATIONS
LINK FOR
REMOTE
PILOTING OR
DATA DOWNLOAD

OPTIMUS

Height	30cm (12in)
Weight	8.5kg (19lb)
Year	2014
Construction material	Composite
Main processor	Commercial processors
Power source	Battery/mains recharge from base station

The quadcopter revolution led by China's Da-Jiang Innovations (DJI) in 2012 (see Mavic Pro, pages 99-101) has opened aerial photography up to everyone. For consumers, this means spectacular vistas of holiday destinations or unusual angles in wedding videos. To commercial users, the range of possibilities is far wider reaching.

Drones can carry out aerial mapping and surveying, tasks that used to require manned aircraft. They can also inspect industrial infrastructure, such as cooling towers and chimneys, without the need to hire a cherry picker or erect scaffolding. Using drones should mean lower costs, but while the drones themselves are cheap, the people operating them are not.

That's why Airobotics Inc. of Tel Aviv have developed a new way to provide drone services without the need for a human operator. Rather than buying a drone, customers lease a 2mt (2.2t) Airbase, which is installed at their site, complete with an Optimus quadcopter. Drone operations are handled remotely by Airobotics.

During operation, the roof of the weatherproof Airbase slides open for the Optimus quadcopter to take off. The drone flies for half an hour and carries daylight or night-time cameras, or special sensors such as LIDAR or chemical sniffers to detect gas leaks. Some missions, such as those for security, might be flown remotely by Airobotics, but many routine tasks can be carried out automatically without human intervention.

The Airbase is roomy – you can get five people in there – and can support multiple drones, though at present only one Optimus flies from each Airbase. The drone connects to the base station to download data, while a robot arm swaps out the battery pack for a fresh one, so there is no delay while it recharges. The same arm can also change the drone's payload, exchanging cameras for other sensors as needed.

Israel Chemicals Ltd is an early adopter of this system, with an Airbase installed at a site in the Negev Desert. It is used to measure stockpiles of phosphate. Previously the stockpiles were surveyed by hand, which involved a surveyor clambering over the heaps, taking measurements using GPS and surveying tools. The site had to be closed to vehicles during this process for safety reasons. Now, the Optimus drone performs this task, flying over the stockpiles on a preprogrammed route each day, shooting high-resolution video. Inside the Airbase, the video is downloaded and passed to Airobotics computers, which turn it into a precise 3D model of the stockpiles via a process known as photogrammetry. Software automatically calculates the stockpile volume and passes a report to the customer. The figures obtained this way are within a few per cent of estimates from manual surveys, and the site stays open throughout.

The biggest challenge in getting this drone-in-a-box solution to work was the automatic landing process. Any misjudgement in a gust of wind could leave the drone crashed and helpless. Airobotics have developed a patented landing technique that they claim provides centimetre accuracy every time – even in changing wind conditions.

Several other companies are working on similar projects for drone-recharging base stations or control centres. Amazon has patented a concept that would have its delivery drones recharging from bases on lampposts and other convenient perches. But Airobotics are the most advanced, and are the first to market. This kind of setup is ideal for regular monitoring. Users are likely to be industrial firms with a need to monitor the condition of pipelines, storage areas and other infrastructure. The drone-in-a-box might also become as essential to security as CCTV, able to arrive swiftly at the scene of an alarm to get a close view of intruders, or even to follow them.

In the longer term, we might see such drones fitted with different sensors so they might perform a range of jobs from monitoring air quality one day, to locating potholes or fly tippers the next and managing traffic the day after. Whatever role they play, it seems that autonomous drones operating from base stations may well become a feature of urban life.

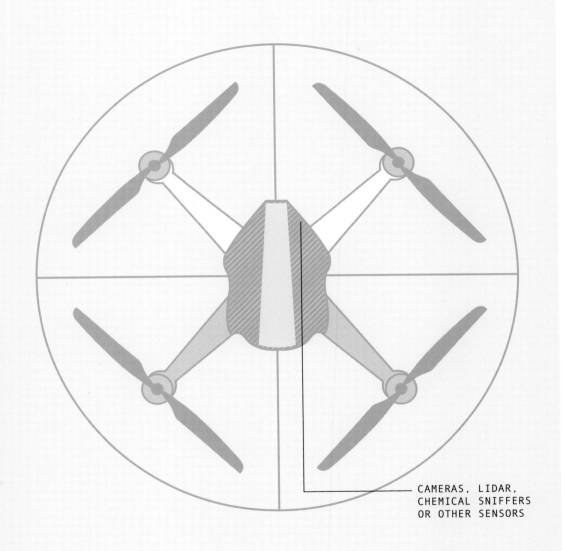

CAMERAS, LIDAR,
CHEMICAL SNIFFERS
OR OTHER SENSORS

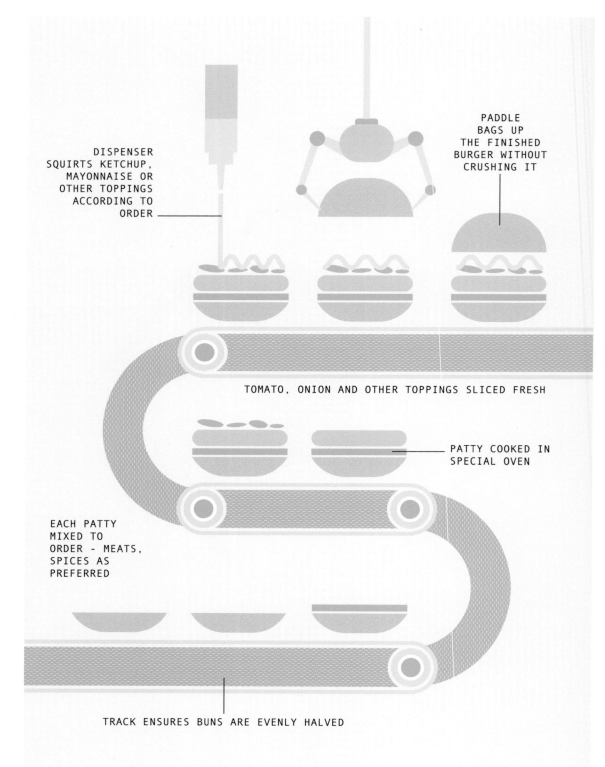

DISPENSER SQUIRTS KETCHUP, MAYONNAISE OR OTHER TOPPINGS ACCORDING TO ORDER

PADDLE BAGS UP THE FINISHED BURGER WITHOUT CRUSHING IT

TOMATO, ONION AND OTHER TOPPINGS SLICED FRESH

PATTY COOKED IN SPECIAL OVEN

EACH PATTY MIXED TO ORDER - MEATS, SPICES AS PREFERRED

TRACK ENSURES BUNS ARE EVENLY HALVED

ALPHA BURGER-BOT

Height	2m (6.6ft), estimated
Weight	200kg (440lb), estimated
Year	2012
Construction material	Steel
Main processor	Commercial processors
Power source	External mains electricity

Fast food restaurants operate on an assembly-line basis. Offering a limited range of items means that orders can go through a fixed series of stages from raw ingredients to ready meal. Given the routine nature of the work, and the cost of employing staff, it was inevitable that entrepreneurs would develop a robot to take over the entire process.

Momentum Machines of San Francisco has developed a robot chef that, they say, does not just flip burgers, but produces gourmet burgers to order. The Alpha is an automated hamburger kitchen in a compact unit occupying just over 2m² (21.5sq ft). It mixes the patties, stamps them into shape and cooks them. Instead of using a griddle it has an oven with a 'secret cooking technique' – this may be a combination of flame and radiant heating for rapid cooking with good results.

Toppings – tomatoes, onions, pickles – are stored in tubes and sliced fresh just before being placed on the burger. The irregular sizes and textures of toppings are a challenge for a machine, making this one of the hardest tasks to automate. 'Cutting tomatoes is a bitch', says Alex Vardakostas, one of the company's cofounders. Vardakostas is an engineer whose family runs a restaurant, typical of Momentum Machines' focus on food and technology. Though buns are easier to deal with, they still require a carefully designed track with sensors and a cutting blade for slicing them and separating the halves. Dispensers squirt precise quantities of ketchup, mayonnaise or other sauces. Finally, the cooked burger, bun and toppings are assembled and bagged. Even the bagging mechanism, which involves a moving paddle to transport the burger into the bag, required considerable thought to ensure that the process was reliable and would never produce a squashed burger.

Alpha produces 360 finished burgers an hour, keeping waiting times to a minimum and serving burgers as fresh as possible. Momentum Machines' goal did not stop at turning out identical products, as MacDonald's does; they wanted every burger to be customised. This could mean a different blend of meat in the patty – for example, having twenty per cent pork or lamb; it could mean a bigger or smaller burger depending on appetite; and the machine is loaded with a selection of speciality cheeses, as well as its wide range of toppings. The company has patented a feedback system to improve service. Customers rate their burgers, and their preferences – say, more cheese and less pickle – are recorded, so Alpha knows what to offer them next time.

Keeping humans out of the kitchen has its advantages. The staff working at the front counter never need to touch any food, so hairnets, gloves and other measures required for hygiene standards can be dropped. Customers need never worry about whether someone in the kitchen is coughing or sneezing and spreading infection, or whether the knife used to slice raw bacon is also chopping onions. Alpha occupies much less space than a normal kitchen, so the dining area can be more spacious. And the developers claim that some of the cost savings from having fewer members of staff will be used to buy better ingredients, to produce 'gourmet-quality' burgers at fast-food prices.

The first version of the Alpha burger-bot was producing burgers to order with ninety-five per cent reliability in 2012. Since that time, the company has kept quiet about developments, but in 2017 they raised venture capital to open their first restaurant, and acquired a location in southern San Francisco.

If Momentum Machines succeeds, other fast-food outlets are likely to follow suit with robot cooks. There will always be a premium on food prepared by a human chef at the high end of the restaurant business, but when it comes to fast food, the priority for consumers is the end product. If there is a demand for customised burgers, or if robots can turn out a better burger for less, then expect to see a lot more robots in the kitchen.

TOPPINGS ARE
STORED IN TUBES
AND SLICED FRESH
JUST BEFORE BEING
PLACED ON THE
BURGER BUN

DISPENSERS
SQUIRT PRECISE
QUANTITIES OF
SAUCE ONTO THE
PATTY

THE BURGER IS
COMPILED ON A
MOVING CONVEYOR
BELT WHICH TAKES
IT THROUGH EACH
OF THE STAGES

ROBOTS
IN YOUR LIFE

ROBOTS IN YOUR LIFE

Seeing a robot in a domestic situation was something of a novelty until recent years, unless it was a toy. From the wind-up tin models of the 1940s that only looked like robots, to Sony's robot dog AIBO in the 2000s, these have become steadily more sophisticated over the years. AIBO had sensors, processors and a computer brain smart enough to learn new tricks and behaviours. It could obey instructions, follow a ball and respond to being stroked. AIBO was the state of the art in electronic toys, but could not do anything useful. Similarly, Lego Mindstorms have more computing power than NASA had when it put a man on the Moon, but they are still toys.

The situation is changing with domestic robots. Millions of people have Roombas and similar self-driving cleaners in their homes, and a growing number also have a robotic lawnmower like the Automower. The iPal goes beyond 'toy' robots, acting as a combined child monitor and entertainment/education device for small children. Millions of older children and adults have a Mavic Pro drone – this might be seen as an expensive item of camera kit or a robotic toy, but either way its capabilities are impressive.

At the hospital, you are increasingly likely to encounter a medical robot like the da Vinci Surgical System for minor surgery, while machines like the Flex Robotic System allow surgeons to perform tricky operations in otherwise inaccessible parts of the body. STAR is an autonomous robot surgeon that will take over the less demanding tasks, such as stitching up a patient. Soft exoskeletons, among them the Wyss Robotic Exosuit, are already helping stroke patients to recover mobility; in future they may become common as aids for active living to the elderly. Some amputees now have sophisticated robot prosthetics, such as the bebionic Hand.

Robots also look set to play a much greater part in everyday life. The Waymo driverless car has received considerable publicity. These robotic vehicles are likely to reshape our cities and their impact may be as great as the invention of the car itself. Meanwhile, Amazon Prime Air's delivery drones might have had a sci-fi edge a decade ago, but their prototypes are already carrying out deliveries.

The Care-O-bot is effectively a robot butler. Although initially aimed at providing help for the elderly, its makers see it as the 'perfect gentleman', always on hand to carry out domestic chores. The Vahana is – very nearly

– a flying car. This electrically powered robotic air taxi for urban environments is able to take off vertically from a small landing pad and carry passengers for a comparable price to a ground taxi.

Not all of these robots will succeed in the real world. Delivery drones and driverless cars – not to mention flying taxis – require changes to existing legislation. Getting the relevant laws changed may take much longer than developing the technology. Simply because a technology is possible does not mean it is desirable. We will need to consider very carefully just how much of a child's upbringing should be left to robots, and how much elderly care and, indeed, how much responsibility in the operating theatre.

Nevertheless, history teaches us that new technology becomes part of the furniture with amazing speed. It shifts seamlessly from novelty to commonplace. Anyone who grows up with a technology finds it completely unremarkable, however astonishing motorcars, televisions or mobile phones might have seemed to their parents a decade before.

Before long it is not a question of having a robot butler, but whether we need one for each member of the family – and how well it can make a soufflé? When everyone has an Automower, people will want robots that can prune trees and shrubs, or plant attractive flower beds. And how about delivery drones for larger items, such as furniture?

Ordinary people in the industrialised West enjoy luxury that was unattainable to any except the rich a few generations before. Machines clean our clothes, hot water arrives at the twist of a tap and our homes are heated at the flick of a switch. Smartphones and the Internet provide what is effectively the sum total of human knowledge in our pockets to carry with us wherever we go. Robots will only extend this process.

We are starting to see robots in our lives. We are likely to see many more.

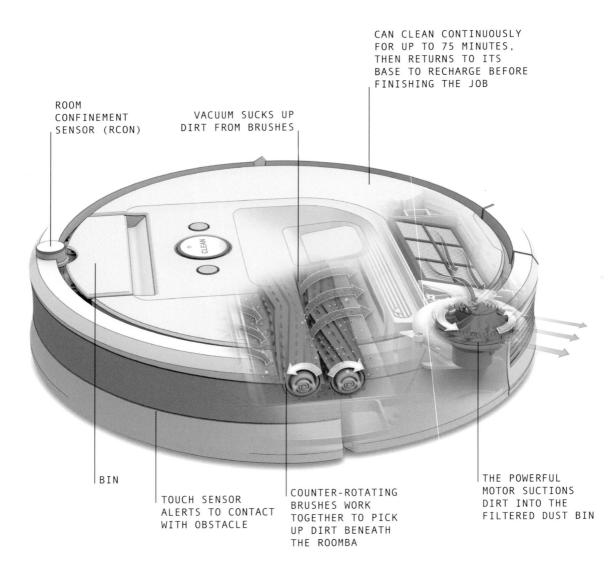

CAN CLEAN CONTINUOUSLY
FOR UP TO 75 MINUTES,
THEN RETURNS TO ITS
BASE TO RECHARGE BEFORE
FINISHING THE JOB

ROOM
CONFINEMENT
SENSOR (RCON)

VACUUM SUCKS UP
DIRT FROM BRUSHES

BIN

TOUCH SENSOR
ALERTS TO CONTACT
WITH OBSTACLE

COUNTER-ROTATING
BRUSHES WORK
TOGETHER TO PICK
UP DIRT BENEATH
THE ROOMBA

THE POWERFUL
MOTOR SUCTIONS
DIRT INTO THE
FILTERED DUST BIN

35CM (1.2FT)

ROOMBA 966

Width	35cm (1.2ft)
Weight	4kg (8.8lb)
Year	2015
Construction material	Composite
Main processor	Classified
Power source	Battery

If housework is not your thing, perhaps you should buy a Roomba. This small, compact device is the first successful domestic cleaning robot, and undoubtedly the most popular robot in the world. More than twenty million Roombas have been sold, all of them working diligently, cleaning floors in homes across dozens of countries.

The original Roomba was launched in 2002 by US company iRobot®. Several generations later, the Roomba 900 series, looks much like its predecessors – a flat disc the size of a dinner plate and about 9cm (3.5in) high. The most noticeable difference, however, is that the machine is about 1kg (2.2lb) heavier than the 2002 original, at 4kg (8.8lb) due to improvements in the vacuum system.

Colin Angle, founder and CEO of iRobot®, originally had a very different vision of what a domestic robot would look like. The first designs had legs for maximum mobility around the house. Angle soon realised that this would not work; legged robots were expensive and unreliable machines, while a cleaning robot needed to be as cheap as other household appliances, and just as dependable. So he reframed the problem: how could a small, cheap robot clean floors? It was not possible to simply miniaturise a vacuum cleaner, so Roomba relies on sophisticated brushwork. Two large rotating rubber brushes lie beneath the machine, while a third, spinning, brush attached at one edge extends to the edges of a room and into corners. All three direct dirt into the path of a vacuum, which sucks it up and stores it in a dustbin.

Roomba trundles along on two large wheels, each driven by its own motor. This arrangement gives the robot a zero turning circle, so it can spin in place to go in any direction. Roomba's low height allows it to scoot beneath furniture and the kickboards in modern kitchens. An infrared sensor notifies Roomba on approaching an obstacle such as a wall, alerting it to slow down. A touch sensor in the front bumper tells Roomba when it runs into something, as which point Roomba repeats a sequence of reversing, rotating and advancing as many times as needed to get past or around whatever is blocking the way. Additional infrared sensors underneath Roomba prevent it from falling off what its makers call 'cliffs' – stairs or similar drops. Should the sensors detect power cords and carpet tassles, Roomba reverses its brushes to avoid tangling. The brushes are not as effective as those of a vacuum cleaner, but the robot can keep going over the floor until the job is done. In fact, the piezoelectric and optical 'dirt sensors' detects the volume of dirt being swept up over a particualr spot to ensure just this. Roomba simply goes over the area again and again.

Originally, models operated using a 'random walk', zigzagging across the room repeatedly until the sensors indicated that it covered the area thoroughly and there was no more dirt to collect. It did so with algorithms originally developed during the Second World War for submarine hunting patrols. The latest Roomba is smarter than that. It maps the room with infrared cameras to figure out where it has been and where it needs to go next. The result is that cleans in straight lines, as a human with a vacuum cleaner would.

When Roomba's battery gets low, it makes its way to a recharging point, which it locates via an infrared beacon. Once recharged, it goes back to cleaning until the job is done. Roombas are typically programmed to clean while their owners are out so they do not get in the way, though their sensors mean that they can work safely around people and pets – you can see how cats and dogs react to Roomba on YouTube.

This robot has obvious limitations. It cannot move furniture or clutter, neither can it clean under cushions or negotiate stairs. This means it cannot do as thorough a job as a human and Roomba owners need to do some vacuuming themselves. Despite these drawbacks, the Roomba is active in millions of peoples' homes – proof that the machine is no longer a novelty, but rather an invaluable aid on the domestic landscape.

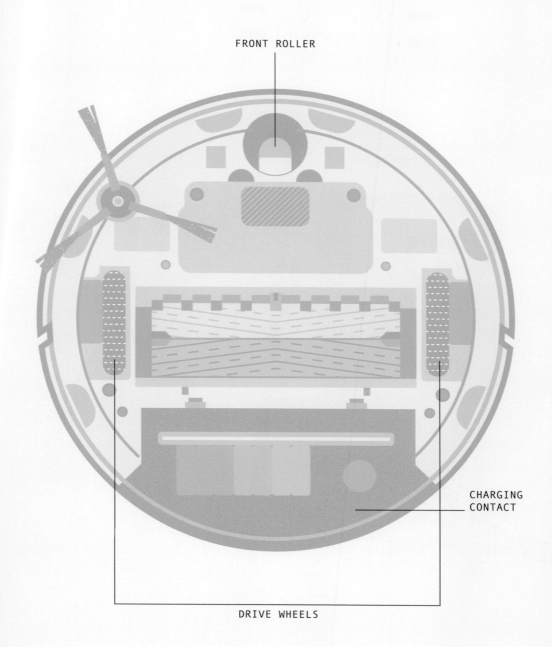

FRONT ROLLER

CHARGING
CONTACT

DRIVE WHEELS

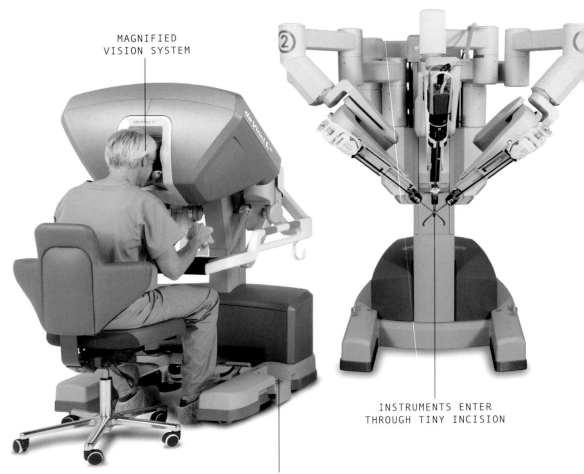

MAGNIFIED
VISION SYSTEM

② INSTRUMENTS ENTER
THROUGH TINY INCISION

SURGEON'S CONSOLE

DA VINCI®
SURGICAL SYSTEM

Height	1.5m (4.9ft)
Weight	544kg (1200lb)
Year	2000
Construction material	Steel
Main processor	Commercial processors
Power source	External mains electricity

Adding a robotic touch to routine laparoscopy, the da Vinci is the world's leading surgical system. Made by US company Intuitive Surgical, Inc., some three thousand da Vinci units are currently installed in hospitals globally. Between them, they have performed over three million operations to date.

The da Vinci is named after Leonardo da Vinci and, specifically, in honour of his creation of what Intuitive Surgical describes as 'the world's first robot'. Theirs is not an autonomous system, but a remote-controlled device that allows a surgeon to operate with greater precision and dexterity than when working manually.

Traditional laparoscopy, or 'keyhole surgery', has been around since the early 1900s, but it only became common in the 1980s. It is a minimally invasive technique, in which operations are carried out deep inside the body via a tiny incision, often less than a 1cm (0.4in) long. The surgeon views the site of the operation via a laparoscope – a camera on the end of a flexible fibre-optic cable – and uses special long-handled surgical instruments to reach the site.

Compared to traditional surgical approaches, laparoscopy means a smaller incision, less blood loss, reduced pain and shorter hospital stays. But it requires considerable skill and dexterity, as there is a reduced range of motion possible to the machine's arm. The instruments enter through a narrow hole that acts as a pivot point, so their ends move in the opposite direction to normal surgery. This 'fulcrum effect' makes laparoscopic surgery challenging to master.

The da Vinci approach, introduced in 2000, makes such surgery simpler with miniature robotic arms. There is a patient-side cart or tower with four robotic arms, and a surgeon's console, usually in the same room as the patient. The magnified vision system resembles those used for normal laparoscopy, but gives a 3D stereoscopic view.

The screen keeps the surgery in front of the surgeon rather than off to one side, and instead of standing, the surgeon sits at the screen with two hand controllers. The robot arms, equipped with a variety of instruments, follow the surgeon's hand movements, automatically smoothing out any hand tremors.

The robot's wrists have seven degrees of freedom, allowing them to be moved into any position inside the incision. This eliminates the fulcrum effect, making it much easier to master. Surgeons can carry out complex tasks, such as stitching and tying knots, more easily inside confined spaces. This reduces the risk of accidents and complications.

The da Vinci system has become the standard for prostatectomy, removal of the prostate gland, and is also increasingly used for cardiac valve repair. It makes difficult operations routine, allowing them to be carried out more quickly and easily.

Perhaps the biggest limitation of the system is its cost; at around $2m, da Vinci is out of the reach of many hospitals. The cost gets passed on, making robot surgery expensive. This makes the system almost unique as, generally, robots are only employed where they are cheaper than humans working alone.

To ease patients' fears of robots running amok, Intuitive Surgical emphasises that 'Your surgeon is 100 per cent in control of the da Vinci Surgical System at all times'. There is also the USFDA (Food and Drug Administration) licensing process to consider; getting the machine certified as safe would be far more difficult were it autonomous.

The Da Vinci Surgical System represents a partnership between humans and robots, much like FANUC's collaborative robots (see pages 11–3). In this case, machines provide the stability and fine control, while humans supply skill and expertise. And the results are better than either could produce working alone.

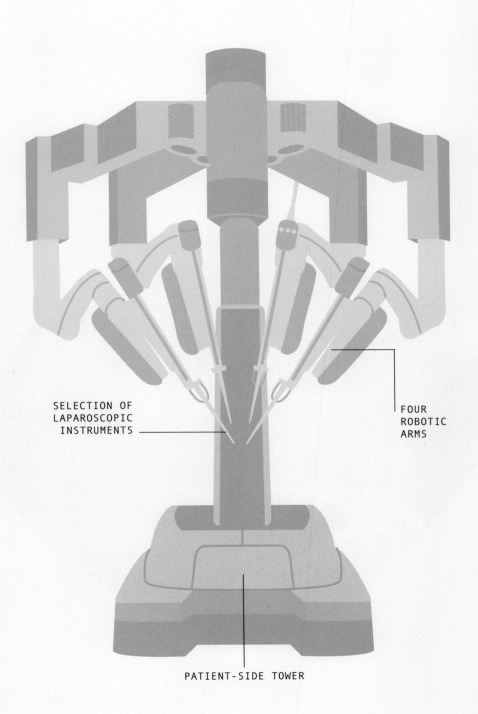

SELECTION OF
LAPAROSCOPIC
INSTRUMENTS

FOUR
ROBOTIC
ARMS

PATIENT-SIDE TOWER

NAVIGATION SENSORS
FOLLOW BOUNDARY WIRES
TO STAY ON THE LAWN

BUMPER
COLLISION
SENSORS

72CM (2.36FT)

AUTOMOWER® 450X

Height	31cm (12.2in)
Weight	14kg (31lb)
Year	1995–2015
Construction material	Composite
Main processor	Commercial processors
Power source	Battery

Mowing the lawn became a far less strenuous task with the advent of powered mowers, but it still required endless walking up and down behind the machine. No wonder that inventors started working on robot mowers from the 1960s, literally as soon as the electronics became available.

Selling its millionth robot in April 2017, the Swedish company Husqvarna entered the world of robot lawnmowers in 1995 and has dominated the market ever since. Its robots are not big, noisy machines with dangerous blades that mow in one single pass. Rather, its Automower takes a gentler approach that works to the robot's strengths.

An Automower mows continuously on a random pattern rather than following the regular up-and-down lines of a human, so the same area may be covered several times. In place of the usual massive blades, an Automower has lightweight safety blades. These exert little force and pivot so that, if they strike anything heavier than grass they just fold away. This makes the robot safe around people. The idea is to 'shave' the lawn, taking off just a few millimetres at a time and mowing daily to keep the lawn at a set height, rather than once a week. Grass cuttings are left on the lawn as natural fertiliser. A sensor detects whether the grass in any one spot is below the height for cutting, so it can tell when the job is done.

The current top-end model is the Automower 450X, which can handle a lawn of up to 0.5ha (1.25 acres).

Like Husqvana's other machines, the Automower is fully automated. The area in which it works is defined by a boundary wire, usually installed by the lawnmower supplier and laid a few centimetres underground by a special cable-laying machine. As the robot approaches this wire, its electric sensor detects a signal in the wire and steers to avoid it. As well as marking the perimeter of the lawn, the wire can mark out islands within it – flowerbeds, ponds or other areas that should not be mowed. The Automower also has a collision sensor for detecting trees, garden furniture and other obstacles. Should the robot collide with something, it stops and changes direction.

The Automower runs on two large wheels, each powered by an independent electric motor. It is highly manoeuvrable and can turn on the spot. Unlike other mowers, it is designed to work in rain – according to Husqvarna, this does not affect its cutting performance. The robot is also designed to be quiet, so it can mow at night without disturbing its owner's sleep. One of the big selling points is that the robot gives you more time to enjoy your garden.

The original 1995 Automower was solar powered, but this proved impractical and they run on batteries. When the battery gets low – after around eighty minutes – the Automower docks with its charging station, either directed by a radio signal or by following the boundary wire or a guide wire. Once set up, an Automower manages the lawn without any human supervision. This means a lawn stays in good shape while owners are away. The operator also set a cutting schedule via a smartphone app so that the robot only mows during the night or while they are at work.

Being an outdoor robot, the Automower has an additional feature – an antitheft alarm system. The alarm sounds when the mower is picked up and cannot be turned off without a PIN. The same PIN is required to reprogram the robot. The 450X also has GPS and geofencing; should it be removed from its designated area, it calls its owner to report itself stolen.

As with many robots, the commercial success of robot mowers has a lot to do with pricing and their ability to compete with cheap human labour. This could explain why, so far, they have been more popular in Europe than in the United States and are generally uncommon at that. However, the continuing success of Husqvarna suggests that, just like grass, the numbers of robot mowers will keep on growing.

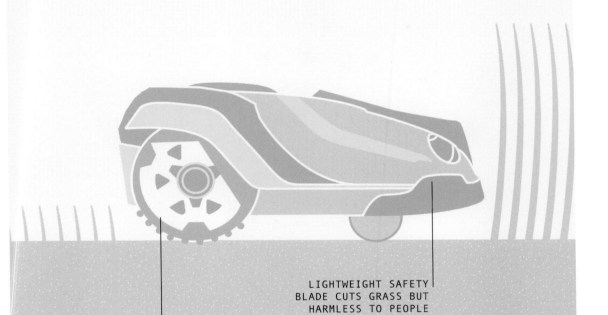

LIGHTWEIGHT SAFETY
BLADE CUTS GRASS BUT
HARMLESS TO PEOPLE

TWO DRIVE WHEELS GIVE
GOOD MANEUVERABILITY

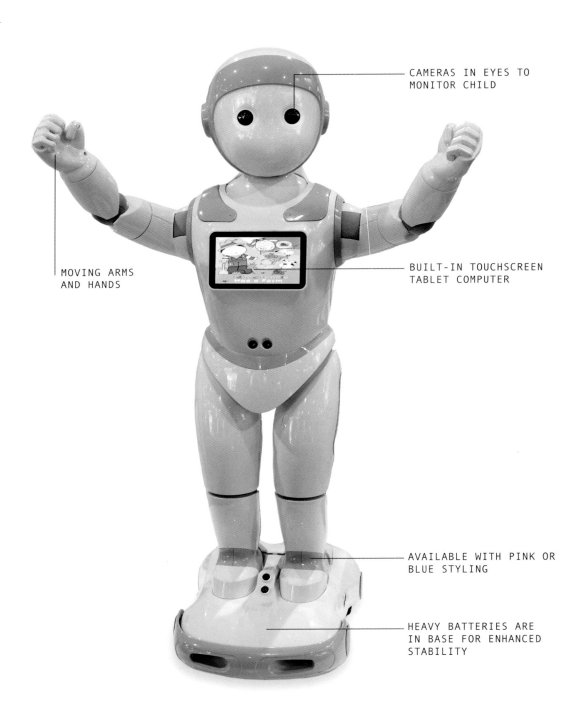

1.06M (3.5FT)

CAMERAS IN EYES TO
MONITOR CHILD

MOVING ARMS
AND HANDS

BUILT-IN TOUCHSCREEN
TABLET COMPUTER

AVAILABLE WITH PINK OR
BLUE STYLING

HEAVY BATTERIES ARE
IN BASE FOR ENHANCED
STABILITY

IPAL™

Height	1.06m (3.5ft)
Weight	13kg (29lb)
Year	2016
Construction material	Composite
Main processor	Commercial processors
Power source	Battery

Tablet computers, which did not exist as consumer items until the introduction of the iPad in 2010, are now firmly established as a part of modern childhood. As of 2017, an estimated one in three children under the age of three in the UK had a tablet, and the average child spends several hours a day in front of a screen. AvatarMind, a Chinese/Silicon Valley startup founded in 2014 want to use a tablet-based approach to help look after children and encourage them to be more active.

The iPal is a 1m (3.2ft) tall robot with big, appealing eyes and a touchscreen tablet in its chest. Built-in sensors allow parents to see and hear their child – each eye houses a 1.3-megapixel camera – and to monitor what a child is doing on the screen. The robot is loaded with educational software; the iPal is so inherently appealing that children simply want to use it. It teaches maths, science and languages, a range that will extend rapidly. It also can keep a continuous record of the child's development, taking pictures or recording video every day.

An iPal runs on four concealed wheels rather than legs, for the usual reasons of economy. Batteries located towards its base give it a low centre of gravity and prevent it toppling over. Collision-avoidance sensors and software keep it from running into things.

The iPal has arms and hands that move, but they are essentially decorative — they cannot grasp or pick up objects, as this was not seen as a useful capability for a child's robot. It can, however, play rock, paper, scissors. The iPal speaks with a squeaky child's voice, and can sing and dance and play games. It has a conversational interface, and the makers say that it can answer a wide range of questions, such as 'why is the sun hot?'. Unlike a human, it never gets tired of answering a stream of questions, or irritated by the same question being repeated, or bored of telling a favourite story again and again.

Not only does the iPal recognise faces, its makers claim that it can pick up on, and respond to, emotions. It also apparently learns the 'preferences and habits' of its host family. Both abilities will improve as the software becomes more sophisticated.

Experts have already raised several concerns about iPal. One is that the robot brings its own agenda. While the makers may not have deliberately programmed the machine to deliver propaganda, it may – like any teacher or nanny – introduce subtle values that are different to those of a child's parents. For example, it may display a tendency to use certain websites or software, or teach creationism rather than evolution.

More seriously, robots like iPal may work too well: children may prefer interaction with the machine to that with other children or parents. Because it spends so much time in their company, this type of robot may actually know more about the child – and on some level, understand them, better – than anyone else. This may lead to 'attachment disorders' in which a child has trouble building relationships in later life.

There is little doubt that such a robot could cause emotional problems if it is used as a substitute for parenting, and the makers are not suggesting that iPal could be a full-time nanny. But robots might be a useful supplement when the parent is working or otherwise unavailable. Plenty of parents buy themselves time with stacks of DVDs or games for iPads; an iPal, or its descendants, seems a more benign alternative, especially if its use is limited to a few hours a day.

Exactly how far such robots could, and should, go remains an open question. Parents would rather teach their own children to swim, or ride a bike, or make pancakes. But parents cannot always be there, especially lone parents. And if a child wants to learn origami, or Swahili, or line dancing, then a patient, seemingly omniscient robot friend such as iPal may give him or her opportunities that might otherwise pass them by.

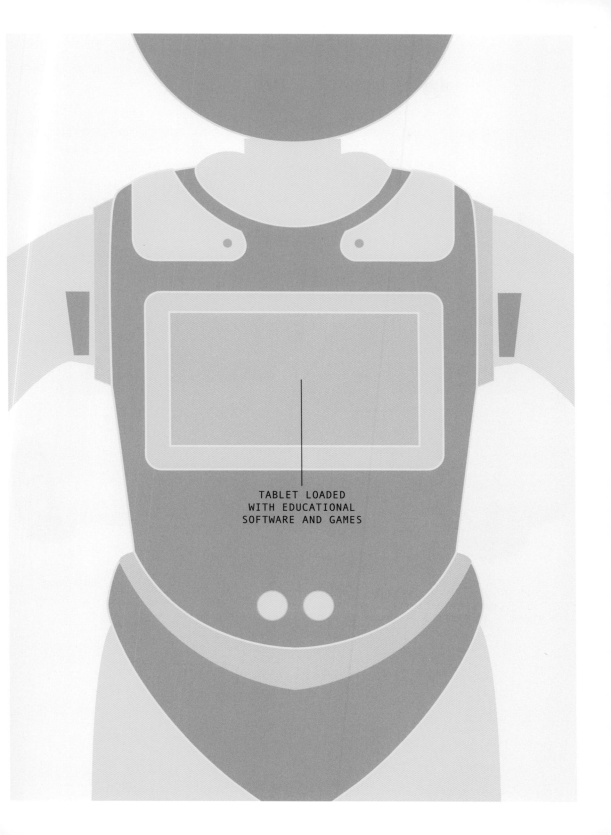

TABLET LOADED
WITH EDUCATIONAL
SOFTWARE AND GAMES

CORNER MOUNTED
RADAR FOR SHORT-
RANGE SENSING

FORWARD-FACING LIDAR
AND CAMERAS

WAYMO TECHNOLOGY INTEGRATED
WITH CHRYSLER PACIFICA
MINIVAN - IT MAY END UP ON A
RANGE OF VEHICLES

CORNER MOUNTED
RADAR

WAYMO

Height	2.1m (6.9ft)
Weight	2.9mt (3.2t)
Year	2016
Construction material	Steel
Main processor	Commercial processors
Power source	Hybrid car: battery/petrol

In 2015, in Austin, Texas, Steve Mahan went for a drive that made history. Mahan is legally blind; but although he was the only person in the car, he was not driving it. Waymo, the self-driving car had carried its first passenger on a public road.

This was a triumph considering the fiasco eleven years earlier, when the Pentagon's Defense Advanced Research Project Agency (DARPA) staged a Grand Challenge for self-driving vehicles. DARPA offered a million-dollar cash prize for the first robot to complete a course through 100 miles (160km) of empty desert. Not one of the fifteen robot contenders finished the course; the best made it no further than 8 miles (13km) before getting stuck. However, robot sensors, processors and software were advancing rapidly. A year later, in the 2005 Grand Challenge, five driverless vehicles made it to the finish. 'Stanley', the winner, was developed by a team at Stanford University led by Sebastian Thrun.

Thrun went on to lead what started as Google's driverless car project, now known as Waymo (both Google and Waymo are subsidiaries of the same parent group, Alphabet Inc.). Stanley had many sensors, but the one that has been most crucial to the development of Waymo is the LIDAR mounted on its roof. LIDAR is a laser-based radar that can be used to build a detailed 3D map of the surrounding environment (see page 44). Google is a data company. One of its key products is Google Maps, in which cities are surveyed and photographed in unprecedented detail. By matching with data from Google Maps, the LIDAR can figure out its exact position and plot its route. In the original Google vehicle, the Velodyne LIDAR accounted for approximately half the $140,000 cost. Google now appears to have developed its own low-cost LIDAR.

Waymo's software is as important as its sensors. The car has to see, understand and react appropriately to everything around it in real time. While most situations require little thought, odd or unexpected events have the potential to cause accidents: unmarked roadworks, a broken-down vehicle or an animal straying into the road.

During its development, Google carried out thousands of hours of testing, with a human 'driver' who kept their hands on the wheel and ensured the robot was driving safely. Google cars were in at least eighteen accidents, all but one of which resulted from human error. In that case, the robot which was pulling out slowly to avoid sand bags placed around a storm drain, was hit by a bus. There were no casualties and only slight damage.

Inside sources suggest that the Waymo car will be electric, but Waymo has not yet revealed what the body of its car will look like. There may even be several versions. Google's test vehicles included Toyota Prius, Audi TT and Lexus RX450h. Google also had a custom vehicle assembled by Roush, a small Detroit-based company that previously worked on projects ranging from aircraft to amusement-park rides. Companies developing rival vehicles include traditional car manufacturers, such as Ford, and tech giants Apple. So far, Google's deep reserves and commitment to self-driving vehicles are keeping it ahead.

Robot vehicles could transform city centres, which are largely designed around cars and car parking. Car parks will cease to be so important when vehicles can park themselves a convenient distance away. Many commentators suspect that the new technology will see car ownership fall off; a driverless car is effectively a robot taxi, so owning one may be less of an issue. Alphabet Inc. has already announced a deal with online taxi company Lyft.

Most cars are only used for an hour or two a day; shared driverless vehicles would mean more efficient usage, making car travel cheaper and reducing the number of parked cars clogging city streets. And when you consider that some ninety per cent of car accidents are caused by driver error, often exacerbated by carelessness, fatigue or alcohol, self-driving cars are ultimately likely to be much safer too.

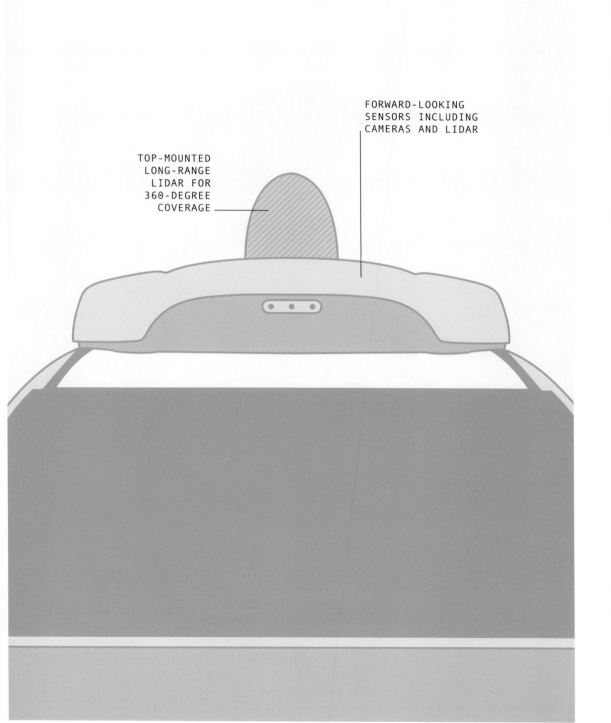

FORWARD-LOOKING
SENSORS INCLUDING
CAMERAS AND LIDAR

TOP-MOUNTED
LONG-RANGE
LIDAR FOR
360-DEGREE
COVERAGE

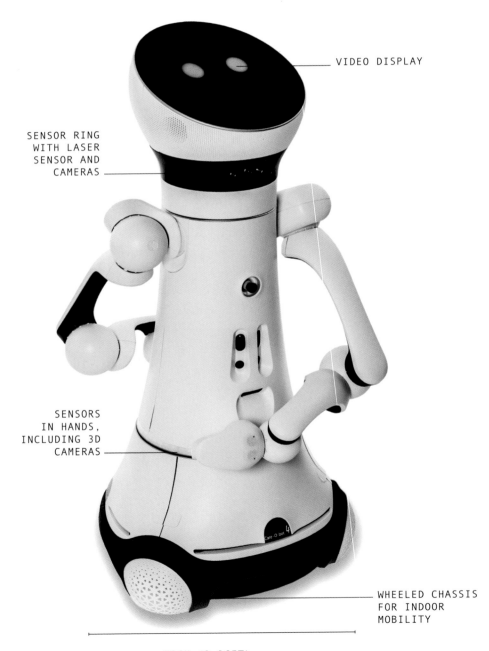

VIDEO DISPLAY

SENSOR RING
WITH LASER
SENSOR AND
CAMERAS

SENSORS
IN HANDS,
INCLUDING 3D
CAMERAS

WHEELED CHASSIS
FOR INDOOR
MOBILITY

72CM (2.36FT)

CARE-O-BOT 4

Height	1.58m (5.18ft)
Weight	140kg (309lb)
Year	2014
Construction material	Composite
Main processor	4-6 Intel NUC i5 256GB, 8GB RAM
Power source	Lithium ion battery

A demographic time bomb is ticking: in the coming decades, there will be more elderly people needing care and fewer young people to care for them. German company Fraunhofer IPA want machines like their Care-O-bot 4 to take some of the load.

Japan has already started promoting elder-care robots, initially aimed at providing interaction and stimulation rather than practical care. PARO is a therapeutic robot that looks like a baby seal, developed for dementia patients; it reduces stress and improves the interaction between patients and human caregivers. Robots like PARO can do some good, as can computers that remind patients about appointments, ensure they take their medication and generally check that they are well and going about their usual activities. But in order to assist human carers more effectively, robots need to be able to help with daily life.

The Care-O-bot 4 is the fourth generation of robots developed by Fraunhofer. It is a mobile, modular testbed for researchers developing caring robots. Its software is 'open source', which means it is easy for users to program it themselves and developers to create their own software.

The simplest variant of the Care-O-bot is a mobile serving trolley that trundles around on four independently steered wheels. A sensor ring has a laser distance sensor and stereo cameras. The robot combines data from the sensors to create a 3D colour picture of its surroundings, enabling it to plan a route and navigate safely, avoiding what the manufacturer calls 'dynamic obstacles' – moving people.

Microphones, speakers, cameras and a video screen enable the robot to function as a 'telepresence platform', a communications terminal that a patient can use to talk to carers, doctors or others without having to master a tablet or other new technology. And of course, carers can use the robot to check on their patient's health.

The Care-O-bot becomes more capable with the addition of one or two arms equipped with spherical joints. Each arm has its own 3D camera, light and laser pointer, and a three-fingered gripper with touch sensors. These allow the robot to adjust its grasping force to pick up objects securely without damaging them. The arm can reach down to pick up objects from the floor or up to high shelves, and it can reach around obstacles without knocking them over.

The Care-O-bot has an adaptive object-recognition system to identify new objects. Once an item is placed in its gripper, the robot rotates it to take pictures from all angles and identifies 'feature points' for recognition and correct orientation. Having been shown an object, the robot can obey requests, such as 'fetch my hairbrush' or 'put the vase on the table'.

Key to the robot's acceptance among elderly users was finding ways in which they could interact with it in a natural, intuitive way. A complicated interface with drop-down menus was not going to work and so the Care-O-bot has voice and gesture recognition. In addition, its torso has two flexible joints that allow it to make body gestures, and a head display to indicate 'mood'. These features make the robot more of a person and less of a thing.

To start with, the Care-O-bot will act as a general butler and domestic helper, able to help with food preparation and serving drinks; ultimately it may do much more. Just like earlier caring robots, it will have a social function as well as a practical one, providing company and assistance.

There is some resistance to the idea of robot carers, owing to the risk that they could increase social isolation. A carer is not just a pair of helping hands, but provides the human touch – empathy, warmth and someone to talk to. The risk is that robots will offer a cheap and easy means to look after elderly people in isolation, leaving them without human contact.

Ideally however, by taking on chores such as cooking and tidying, robots like the Care-O-bot 4 will free up time for human carers who can then concentrate on the caring part of their job. And by providing assistance – and a means of connecting to the Internet and to the world – they will improve the quality of life of elderly people.

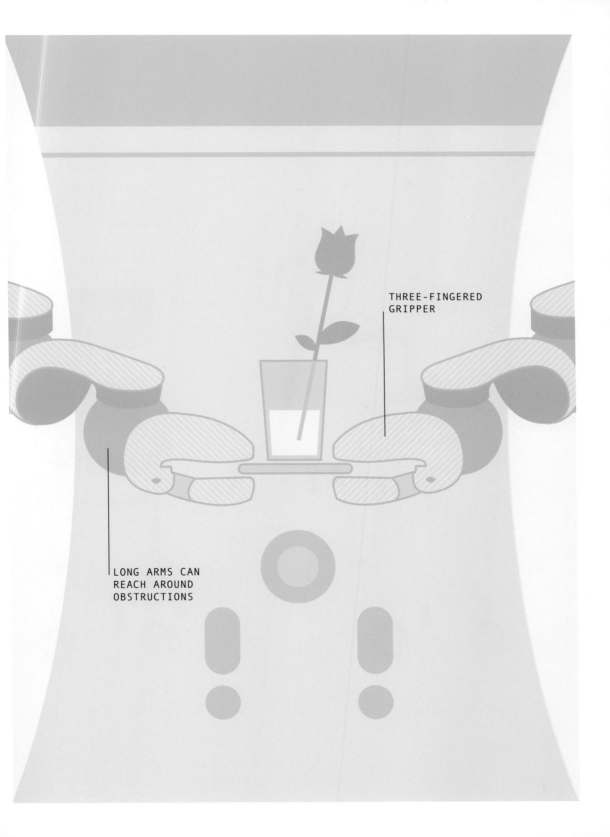

THREE-FINGERED
GRIPPER

LONG ARMS CAN
REACH AROUND
OBSTRUCTIONS

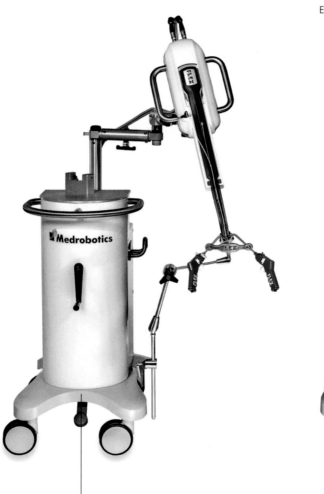

CONTROLS TO STEER
ENDOSCOPE AND OPERATE
SURGICAL TOOLS

PATIENT-SIDE CART

SURGEON'S CONSOLE

FLEX® ROBOTIC SYSTEM

Height	136cm (4.5ft)
Weight	196Kg (430lbs)
Year	2014
Construction material	Steel
Main processor	Commercial processors
Power source	External mains electricity

Robots such as the da Vinci Surgical System (see pages 67–9) still enter the body via a surgical incision, even if the incision is as small as possible. But there is another way: accessing the interior through natural orifices. This technique, called endoscopy, is well established for imaging. The science of robotics is extending this approach further, with devices that can crawl into the human body and carry out operations.

Philipp Bozzini invented the first rigid endoscope in 1806. His Lichtleiter, or 'light conductor', allowed him to look into patients' ears, nasal cavities and rectums without a need for surgery. Later physicians developed flexible endoscopes that could travel further into the body. They then added a needle for taking tissue samples and there is now a growing range of surgical tools that can be fitted to an endoscope.

When Professor Howie Choset of the Robotics Institute at Carnegie Mellon University developed the Flex Robotic System, he took endoscopy to the next level. Medrobotics Corp of Raynham in Massachusetts launched Choset's machine in 2014. As with da Vinci, the system has two components: the robot itself, which is positioned beside the patient, and a surgeon's control console, normally be in the same room.

A flexible robot travels down a patient's throat, under the control of a surgeon using a joystick, in turn guided by the magnified image from a video camera in the robot's nose. The robot is a mechanical snake, consisting of a series of articulated links capable of flexing around 180 degrees. Powered by four motors, the number of joints give Flex an impressive 102 degrees of freedom.

When the surgeon reaches the site of interest, the articulated sections are locked in place and the robot becomes a stable operating platform. Two small tubes inside the endoscope allow surgical instruments to be fed to the site. These include scalpels and scissors, grippers and a needle able to stitch up wounds. A surgeon using the video camera can use grippers to hold a piece of tissue and then cut it with the scalpel.

The makers of Flex say that the robot's high level of manoeuvrability within the body allows surgeons to carry out operations in places that would previously have been difficult or impossible to reach without an incision. Following initial US trials, in 2016, Flex had a ninety-four per cent success rate in reaching the site of a tumour, with fifty-eight per cent of those sites being classified as 'difficult to reach' by conventional methods.

The Flex represents a new level of minimally invasive surgery and, like the da Vinci, it is very much a remote-controlled system rather than an autonomous one. The USFDA licensing process required that a surgeon be in control at all times, otherwise the machine would not have been certified as safe to use. Safety features include a controller to ensure that there are no accidental movements. This prevents a robot from progressing unless the surgeon has a foot on a pedal. The only truly autonomous feature is retraction: following an operation, the Flex automatically withdraws back the same way it came.

Professor Choset sees his Flex robot as the start of a new type of healthcare. Unlike the da Vinci, which tends to be confined to elite medical institutions, the Flex is highly affordable. Minimally invasive surgery that is carried out inside the patient does not necessarily require a hospital stay – or even a hospital. For routine operations, Choset believes that surgical procedures might be delegated to non-surgeons. Such a combination might lead to what Choset calls the 'democratisation' of surgery, with operations being carried out far more quickly and easily, with shorter waiting lists and far fewer patients even visiting a hospital.

With the cost and complexity of medical treatment steadily spiralling upwards, robots such as Flex may start to reverse the trend. They may not be welcomed by everyone in the medical industry, especially well-paid surgeons who stand to lose out. But if flexible robots can safely provide better outcomes at lower costs, there is no reason why they should not become more widely as surgical tools in time.

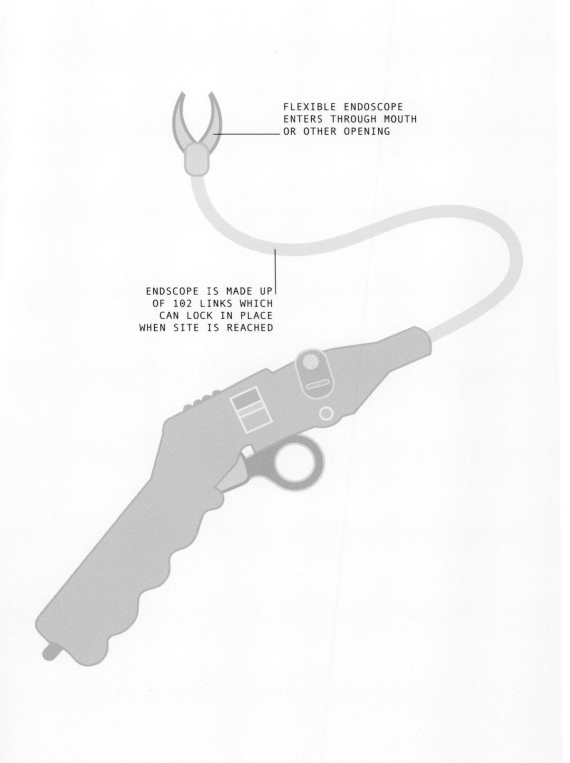

FLEXIBLE ENDOSCOPE
ENTERS THROUGH MOUTH
OR OTHER OPENING

ENDSCOPE IS MADE UP
OF 102 LINKS WHICH
CAN LOCK IN PLACE
WHEN SITE IS REACHED

DELIVERS PACKAGE OF
UP TO 2.2KG (4.9LB)

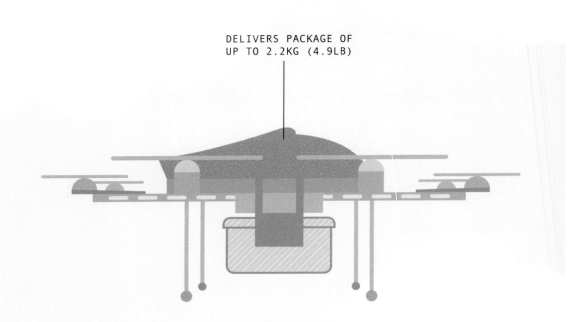

EACH CUSTOMER'S LANDING MAT
HAS A UNIQUE ID SO DRONE DELIVERS
TO THE CORRECT SPOT

AMAZON PRIME AIR

Height	30cm (12in) estimated
Weight	6kg (13.2lb) estimated
Year	2013
Construction material	Composite
Main processor	Commercial processors
Power source	Battery

Amazon is by far the world's most successful online retailer. Customers can order from more than one hundred million items, and a giant warehouse-cum-delivery centre usually ensures that goods are delivered the next day. The company's growth has been fuelled by aggressive innovation – using Kiva robots for 'picking and packing', for example (see pages 27-9). But when the company's Christmas 2013 advert featured a delivery drone, it looked more like a publicity stunt than a real plan.

Initial, frivolous demonstrations had shown that it was technically possible for a quadcopter to deliver pizza, beer or related items, but the idea had never been presented as serious business proposition. While the Amazon advert seemed to be an amusing attention-grabber, the company did indeed hire teams of drone developers, set up testing sites, and lobby for changes in the air regulations to permit commercial drone deliveries.

Not yet up and running, Amazon Prime Air, will not fulfil all of Amazon's orders. It will be a premium service for items needed urgently, delivering packages of less than 2.2kg (4.8lb) within 30 minutes. This will require an automated setup in which drones fly themselves without human input.

The 2013 advert gave some idea of Amazon's vision. The drone is battery-driven, as are consumer quadcopters, but is of a conversion type. This means it takes off and lands vertically, like a helicopter with rotors, but in the air, it is driven forwards horizontally by a propeller, like an aeroplane.

This allows the drone to fly at up to 60mph (96kmh) for a distance of 8 miles (13km), and gives it an essential vertical-landing capability. In the video, the customer lays out a landing mat, and the drone gently sets down the delivery upon it. Landing pads may have a unique identifier, such as a barcode, so the drone can pick out the right one when there are several landing mats in a small area.

The first 'real' Amazon drone delivery was carried out in December 2016 in the UK, where a small-scale trial of the service is underway. The order consisted of a TV streaming stick and a bag of popcorn; again, it looked like a publicity stunt. The drone itself was very different to the 2013 version, being a quadcopter with rotors protected by a plastic guard so they are not a safety hazard.

The big issue with this sort of delivery is that drones must be able to fly safely over densely populated urban areas. At present, drones are not allowed to fly out of sight without human control. Amazon is working on a sense-and-avoid system using cameras and other sensors to spot obstacles including trees, buildings, wires, birds and other drones. Amazon's plan is that its drones will operate in the airspace below 400ft (120m), keeping them out of the way of manned aircraft. Aviation authorities are still considering how to regulate such activities: drones will not only need to be safe, they need to be proven to be safe. Even a drone weighing a few kilos could cause severe injury at 60mph (96kmh).

The United States is likely to be one of the toughest areas to get drone deliveries approved, so Amazon has pilot projects in other places. A working system in one country may help pave the way for regulations elsewhere.

In the meantime, Amazon has registered a steady stream of patents relating to Amazon Prime Air, on ideas as diverse as airships acting as drone base stations, to recharging stations on skyscrapers and orders delivered by parachute. The drones themselves may look different, but the drive towards a drone-based service seems unwavering.

Many other companies are working on similar concepts. These range from Matternet, who intend to set up a drone network to deliver drugs and medical supplies to remote villages in developing countries, to traditional delivery companies such as DHL and UPS experimenting with drones flying from delivery trucks. Internet giant Google has a rival delivery drone project known as 'Project Wing'.

Such a broad overall level of investment so far suggests a strong belief in delivery drones. Amazon may be the first to get their network going, but within a few years drone deliveries could be as commonplace as emails.

VERTICAL PROPELLER
FOR HIGH-SPEED
FORWARD FLIGHT

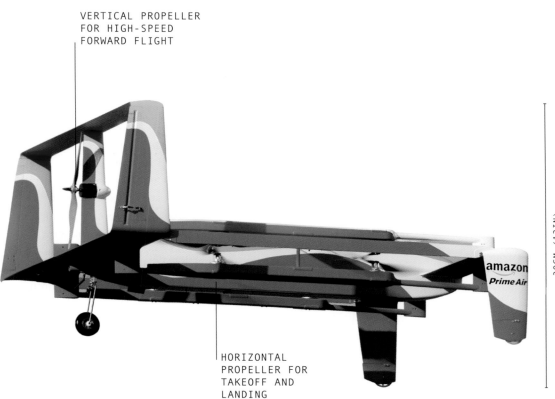

HORIZONTAL
PROPELLER FOR
TAKEOFF AND
LANDING

30CM (12IN)

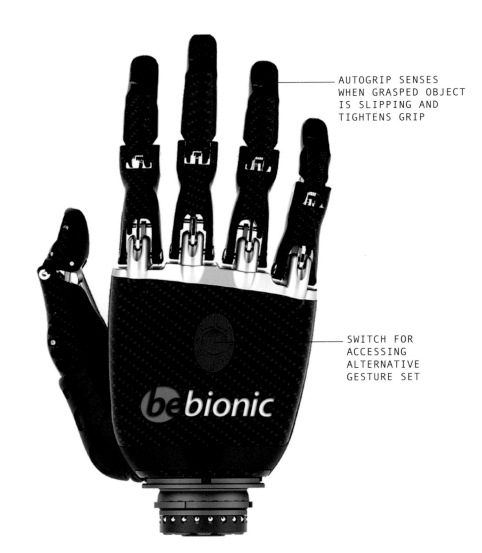

19CM (7.5IN) MEDIUM SIZE/20CM (8IN) LARGE SIZE

AUTOGRIP SENSES
WHEN GRASPED OBJECT
IS SLIPPING AND
TIGHTENS GRIP

SWITCH FOR
ACCESSING
ALTERNATIVE
GESTURE SET

be·bionic

USUALLY WORN WITH A
SILICONE GLOVE MATCHING
THE WEARER'S SKIN TONE

BEBIONIC HAND

Diameter	5cm (2in) diameter at wrist
Weight	598g (1.3lb)
Year	2010
Construction material	Aluminium alloy
Main processor	Commercial processors
Power source	Battery

The most direct application of Leonardo da Vinci's concept of machines imitating human articulation lies in prosthetics. Replacement limbs have evolved from simple hooks or lifelike, but useless, wooden models to sophisticated robotic devices able to match the actions and movements of a human hand.

UK company RSL Steeper claim that the bebionic hand it developed is the most advanced prosthetic hand in the world. Wearers can carry out a wide range of activities that the rest of us take for granted, but that can be challenging for amputees: picking up a glass, turning a key in a lock or carrying a bag, for example.

The term bionic is a combination of 'bio' (living) and 'ic' (like, or having the qualities of) and was popularised by the TV series, *The Six Million Dollar Man*, about a test pilot with limbs replaced with super-powered prosthetics.

The bebionic hand is battery-powered, and controlled directly by the user's own muscles. Each of its five digits has its own independent motor. An electronic sensor detects changes in the skin conductivity over control muscles in the wearer's wrist, and by working these muscles the wearer activates a grip. This 'myoelectric' control has been used for some years, but what distinguishes bebionic is the level of control that it provides with the different grip options it offers. There are fourteen basic hand positions or grip patterns for different tasks.

One set of grip patterns is available directly, and users can access another set by pressing a switch on the back of the hand. A user selects their own primary and secondary grip patterns based on their preferences and needs. The manufacturers say that operating the hand becomes instinctive after some training and practice.

What makes the hand genuinely robotic is that it has its own intelligence: embedded microprocessors monitor the position of each finger for precise, reliable control. An 'auto grip' function senses when something is slipping out of the hand's grasp and adjusts the grip to keep hold of it. For some grips, it is necessary to shift the thumb manually between the opposed and non-opposed position. The opposed position is used in grips such as the 'tripod' – for holding things like pens – in which the index and middle fingers meet the thumb. An opposed thumb is also used for the 'power' grip where all four fingers close towards the thumb – this is the grip you use to pick up a ball or a piece of fruit, or cylindrical objects such as bottles, glasses or kitchen and garden tools.

Non-opposed hand positions include the 'finger point', where the middle, ring, and little fingers are folded into the palm and the thumb rests on the middle finger, with the index finger partially extended. This is the position used for typing – using just the index finger on each hand – or pressing buttons. The 'mouse' grip, as the name suggests, is handy for operating a computer mouse, while 'open palm' is useful for carrying large objects, such as trays or boxes. The hand can carry weights of up to 45kg (99lb) and is strong enough to crush a can. Proportional speed control ensures the right level of force; as well as crushing cans the user can pick up something as delicate as an egg without breaking it.

The bare hand looks robotic, but is covered in a removable silicone glove, which comes in nineteen different shades to match a user's skin tone, complete with customised silicone nails. Users may tend to reject prosthetics for failing to feel like a part of themselves, but bebionic has overcome the resistance of many with its strong combination of looks and functionality.

It may not provide super-strength – a notable feature of the original Bionic Man – but the bebionic hand is improving the daily lives of its users in ways never seen before .

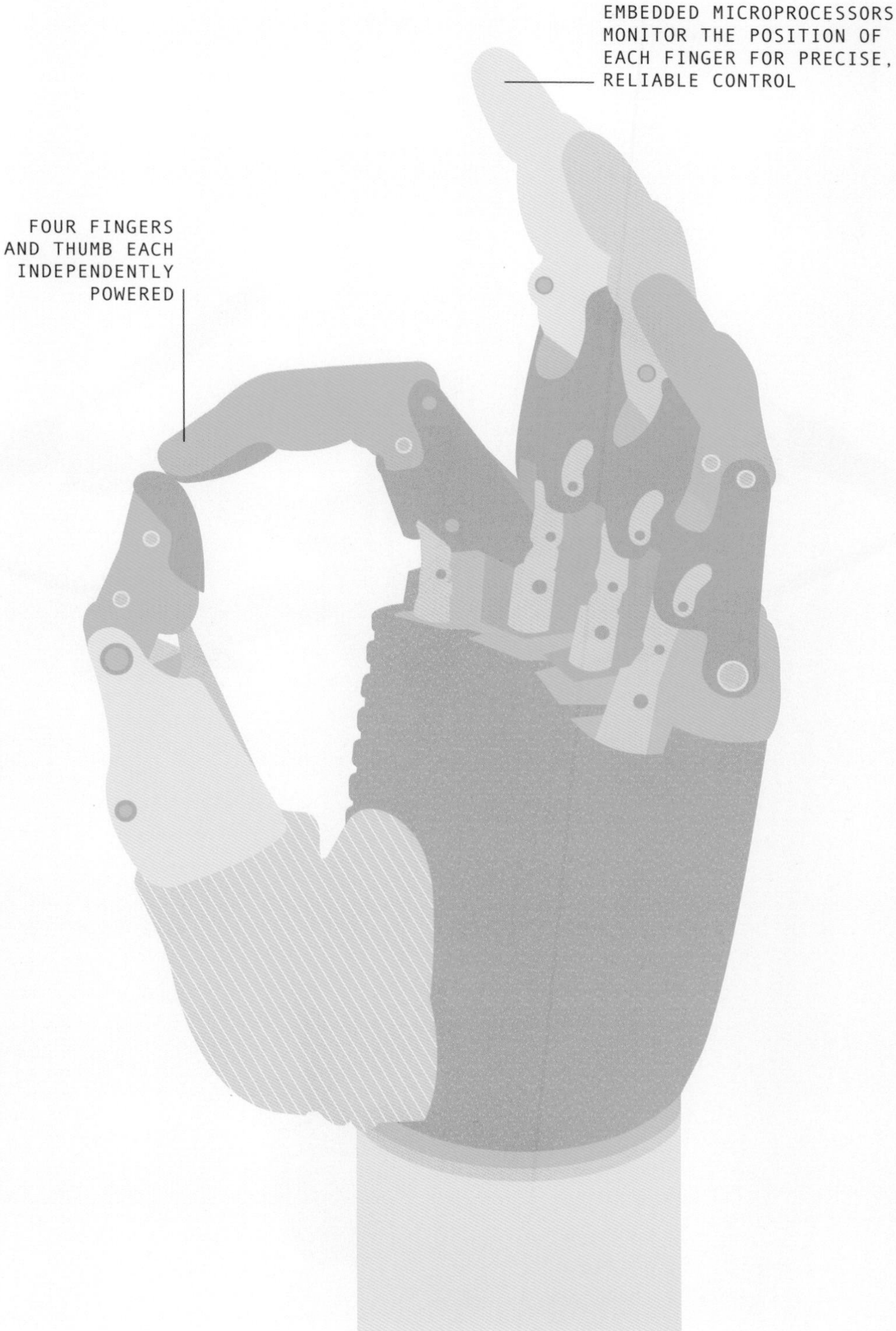

EMBEDDED MICROPROCESSORS
MONITOR THE POSITION OF
EACH FINGER FOR PRECISE,
RELIABLE CONTROL

FOUR FINGERS
AND THUMB EACH
INDEPENDENTLY
POWERED

FOUR SETS OF
BLADES WITH
SEPARATE MOTORS
STEER AND MAINTAIN
STABILITY

STABILISED 4K CAMERA

WITH THE WINGS EXPANDED, THE MAVIC MEASURES 198MM (7.8IN) IN WIDTH

THE USER INDICATES
WHERE THE MAVIC
NEEDS TO GO TO ON A
TOUCHSCREEN MAP

MAVIC PRO

Height	8.3cm (3.25in)
Weight	743g (1.63lb)
Year	2016
Construction material	Plastic
Main processor	Proprietary processors
Power source	Lithium ion battery

Miniature quadcopters, such as the Mavic Pro, are an increasingly common sight. Millions of enthusiasts use them to shoot spectacular aerial videos, and their popularity has made Chinese makers Da-Jiang Innovations (DJI) the most successful drone manufacturer in the world. While the Mavic Pro may look like a toy, closer inspection reveals a highly sophisticated autonomous flying robot.

Multicopters can be traced back to US engineer Mike Dammarm, who developed his first battery-powered quadcopter in the early 1990s. Standard helicopters, with a horizontal rotor blade for lift and a vertical rotor for stability. They steer by changing the angle or pitch of their blades, which requires an elaborate mechanical arrangement. The quadrotor is mechanically far simpler. Four sets of blades, each with separate motors, steer and maintain stability by speeding up or slowing down different rotors. This would be impossible for a human pilot to control, but electronics handle the task with ease. The Mavic Pro can remain perfectly stationary in the air, even in changing wind and weather conditions. It can also tilt to fly forward at high speed.

Dammarm's invention hit the big time in 2010, when French company Parrot introduced the AR.Drone, a quadcopter that sent back video in real time via Wi-Fi. The AR.Drone was a bestseller, but could only fly for a few minutes at short distances, and the camera had poor photo resolution.

Frank Wang, DJI's CEO, saw the potential of small drones as platforms for aerial photography, and set about developing the technology – batteries, motors, cameras, sensors and computer brain – to realise that potential. In 2013, the company launched the Phantom, which could take high-quality video for twenty minutes at a stretch, and could be flown from 1 mile (1.6km) away. The Phantom was priced for the mass market and DJI sold them by the million.

The new Mavic Pro is smaller than the Phantom series, weighing just over 700g (1.6lb) and priced at less than £1,000, but is packed with advanced electronics. The manufacturer claims that it is simple enough that a beginner can unpack the machine and start flying it at once. The Mavic Pro flies at 40mph (65kmh), shooting 'movie-quality' 4K video for up to twenty-five minutes with a stabilised camera. It can be controlled from 5 miles (8km) away.

Despite its small size, the Mavic copes with winds of over 20mph (32kmh). Downward-facing sensors – cameras and sonar – allow it to hover as steady as a rock, and to maintain a constant level above ground as it mounts a slope. Portability is a key selling point. The Mavic Pro's arms fold away, making it the size of a water bottle, so climbers or hikers can pack it safely and easily unfold it for action.

What makes the Mavic Pro a robot is the high level of built-in automatic control. It can fly itself, and avoids obstacles with the aid of sonar and cameras. The user indicates a destination on a touchscreen map, and the drone navigates the route itself via GPS. The Mavic Pro can always return to its takeoff point at the touch of a button. A clever optical navigation system takes pictures of the ground below at takeoff and uses these to recognise the launch point, enabling it to return and land automatically a few centimetres from where it started.

The Mavic also has a tracking mode: the camera locks on to a person or vehicle and follows them. This means that skiers, skateboarders and others can capture video of themselves in action. Gesture mode is another way of directing the drone without a radio-controlled unit – the drone follows hand signals given by the operator, taking pictures when requested.

The Mavic Pro's larger cousins have revolutionised professional film-making. You no longer need a helicopter to get aerial shots of a car chase, to swoop into a canyon or over a skyscraper, or to fly with eagles. Machines like Airobotics' Optimus drone-in-a-box (see pages 51–3) are moving into industry and agriculture. But, on a domestic level, the robot that most people are likely to own and enjoy will be a Mavic Pro.

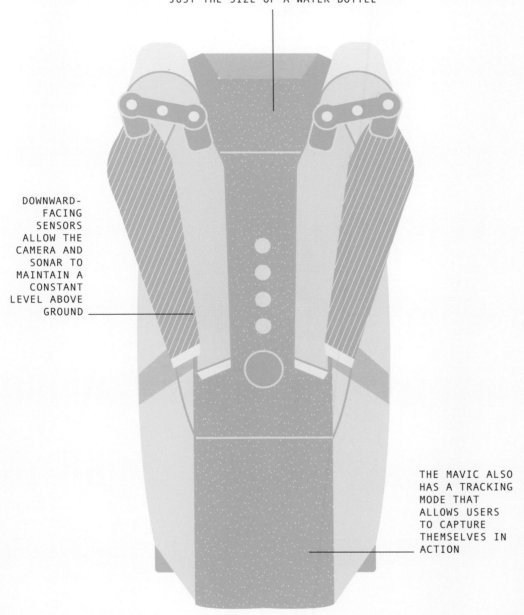

THE MAVIC'S 'ARMS' FOLD BACK
FOR EASY TRANSPORT, MAKING IT
JUST THE SIZE OF A WATER BOTTLE

DOWNWARD-
FACING
SENSORS
ALLOW THE
CAMERA AND
SONAR TO
MAINTAIN A
CONSTANT
LEVEL ABOVE
GROUND

THE MAVIC ALSO
HAS A TRACKING
MODE THAT
ALLOWS USERS
TO CAPTURE
THEMSELVES IN
ACTION

HEIGHT: DEPENDS ON WEARER

SENSORS MONITOR
WEARER'S MOVEMENT

SOFT ROBOTIC EXOSUIT

Height	1m (3.3ft) approx.
Weight	3.5kg (7.7lb).
Year	2014
Construction material	Composite
Main processor	Commercial processors
Power source	Battery

An exoskeleton is a wearable robot that matches its wearer's movements and gives them additional strength. While the initial motivation for this type of work came from the military's desire for armoured suits resembling Iron Man, with superhuman strength (see XOS 2, pages 123–6), researchers are now looking at a different type of exoskeleton for use on a domestic level.

Soft exoskeletons do not form a rigid shell, but act as an additional set of muscles. Conor Walsh leads a team at Harvard University's Wyss Institute for Biologically Inspired Engineering, developing a range of devices based on this concept. They provide mobility assistance to patients weakened by strokes or other conditions, and can make it easier for people to carry heavy loads or walk long distances. Like Leonardo da Vinci, the researchers have had to look closely at how human muscles and tendons work.

Walsh's 'lower-extremity robotic exosuit' fits over the wearer's legs, with a waist belt carrying motors, pulleys and a battery pack. Two vertical straps on each leg connect the waist unit to the ankles. Sensors monitor the wearer's movements and a computer system determines the right moment to activate each of the motors. The exosuit provides assistance precisely when help is needed to take a step, lift a foot, push off the ground or place a foot down, adjusting itself seamlessly to match the wearer's gait. The result of such sophisticated control is that the exosuit does not seem to be doing anything; the wearer only notices that it easier for them to walk.

Instead of relying on electric motors, the exosuit uses a type of artificial muscle known as a McKibben actuator. These are pneumatic, powered by compressed gas, and behave like real muscle. Unlike electric motors, muscles are not consistent in the amount of force they exert, which depends on how far they are extended. To work in concert with human muscles, the artificial version needs to have a similar pattern of strength, otherwise it would push too hard at the beginning and end of each movement with too little in the middle, interfering with, rather than assisting movement. When deactivated, the soft exoskeleton is just a heavy additional garment; the overall weight is significant at 3.5kg (7.7lb).

Stroke patients have been among the first to use exosuits. Eighty per cent of stroke survivors lose some function in one limb, and an exosuit can help them to regain their walking ability. Typically, stroke patients change the way they walk, lifting their hips or swinging a foot around in a circular motion to prevent it dragging on the ground. Such measures solve the immediate problem, but limit mobility. An exosuit helps them walk normally and easily, preventing them from developing a poor gait during rehabilitation.

Walsh's team has also developed a soft robotic glove for patients who have lost strength in their fingers. Like the lower-limb exosuit, this works with the wearer's body and effectively amplifies their remaining strength. In the future, electromyography sensors could detect nerve signals when a wearer attempts to carry out a movement such as grasping, and the glove will function even if the patient is not strong enough to move their fingers. This could be extremely useful for patients not able to carry out basic tasks, such as picking up a cup or using a fork, without assistance.

Another project is a lower-limb exosuit for hiking for extended periods carrying heavy loads. The military are obvious customers. Tests have shown that the soft exosuit decreases the 'metabolic cost' of jogging and other activity. So, for example, it takes less effort to walk 1 mile (1.6km) even with the added weight of the exosuit.

Soft robotic exoskeletons are being trialled for commercial use. In the United States, Retailer Lowe's has experimented with shelf-stackers wearing soft exoskeletons to reduce the risk of back injury.

Soft exoskeletons are still in the development stage and, going forward, it will be a matter of improving the them in order to increase the overall benefit. But as the technology develops they may become increasingly common,allowing people to enjoy travel and outdoor activities long into their later years.

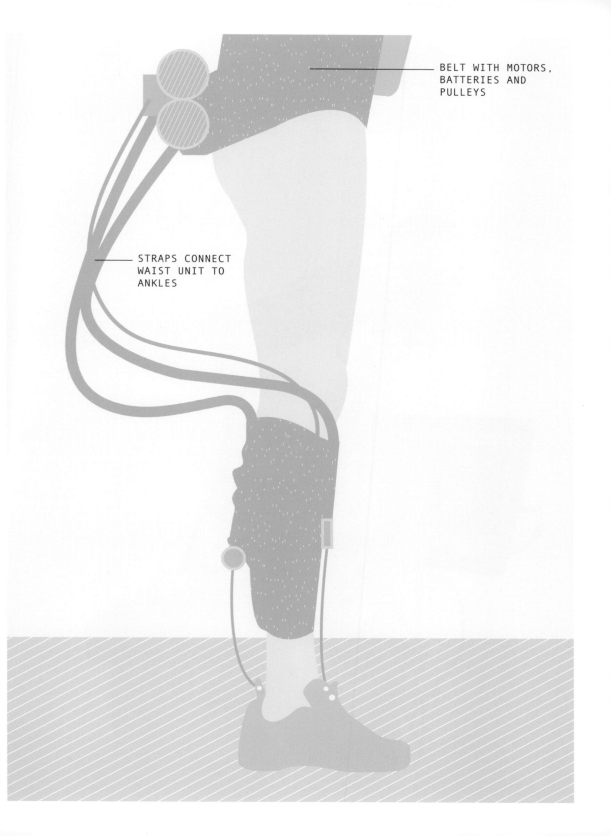

BELT WITH MOTORS,
BATTERIES AND
PULLEYS

STRAPS CONNECT
WAIST UNIT TO
ANKLES

1.5M (4.9FT)

COMMERCIAL
ROBOT ARM

MAGNIFIED
SURGEON'S
DISPLAY

SUTURING
TOOL

SMART TISSUE AUTONOMOUS ROBOT

Height	1.5m (4.9ft) approx.
Weight	30kg (66lb).
Year	2016
Construction material	Steel
Main processor	Commercial processors
Power source	External mains electricity

Robots such as the da Vinci Surgical System and the Flex Robotic System robot (see pages 67–9 and 87–9) have to be controlled by a surgeon because of the requirements of soft-tissue surgery. Automating the process calls for a new type of robot altogether.

If the subject stays in a fixed position, an industrial robot can do the job of the surgeon – it is just a matter of drilling in a certain place, attaching a screw, and so on. For knee operations and laser eye surgery, this type of automation is routine, though it is rarely described as 'robotic' surgery, possibly for the patient's peace of mind.

Soft-tissue surgery is more difficult. The squishy stuff that makes up most of our bodies is apt to move around during a surgical process, so a robot cannot simply be programmed to carry out the sort of fixed moves used in bone surgery. Peter Kim and colleagues at the Sheikh Zayed Institute for Paediatric Surgical Innovation in Washington DC designed the Smart Tissue Anastomosis Robot (STAR) to tackle this challenge. STAR's special skill is tracking the movement of tissue in three dimensions.

Kim borrowed technology from the movie business. 'Motion capture' is a technique for filming an actor's movements in 3D so they can be overlaid onscreen with computer-generated imagery – for example, Andy Serkis' performance as the grotesque Gollum in the Hobbit series. During motion capture, a series of fluorescent dots are positioned on the actor's bodysuit at key locations – in particular the joints – to show the relative movement of hands and elbows, knees and feet. Multiple cameras track the movements of these dots from different angles. A computer then creates a wireframe animation of the action based on the dots, which forms the scaffolding for the CGI. STAR uses a similar technique. Before surgery, a series of Near Infrared Fluorescent (NIRF) tags are placed around the operation site. A set of 3D cameras follows the movement of the tags tracking the precise position of the tissue.

STAR was initially set up to stitch or suture a wound. A robotic arm with seven degrees of freedom from German company, Kuka AG, is fitted with a special stitching device. This has a curved needle that arcs round, putting in the stitches. A sensor detects the level of force being applied by the needle to ensure it pushes hard enough to go through the skin, but not so hard as to cause tears.

STAR has been tested on dead tissue and on animals, and it has performed well – better than human surgeons, according to its developers. They also claim that it can stitch faster, with more consistent results. The stitching leaked less, and fewer stitches needed to be redone, than with a typical human surgeon. At the moment STAR only works under close supervision, but in principle it should be possible for a doctor to mark out a line that needs stitching, tap a button and have STAR do the work.

STAR is not a prototype, but a proof-of-principle of what robots are now capable of. One surgeon described this type of suturing as a 'grand challenge', the type of task that needs to be mastered before robots can progress. Suturing is a small but important start. If it can be fully mastered and shown to work safely, then the universe of soft-tissue surgery could open up for STAR. In principle, machines such as the da Vinci Surgical System and the Flex Robotic System could be upgraded to become fully autonomous robot surgeons.

Initially, this technology may be adopted in situations where no human surgeon is available, for example in space missions. Such missions usually include a doctor, but sometimes the doctor is also a patient. In 1961, Leonid Rogozov, the doctor at a Russian Antarctic base, removed his own appendix with the aid of a mirror, anaesthetic and several unqualified assistants.

Looking further forward, surgical robots are likely to become increasingly capable. Ultimately, they may even be able to carry out operations of a delicacy and complexity that currently prove impossible for human surgeons, such as some forms of neurosurgery.

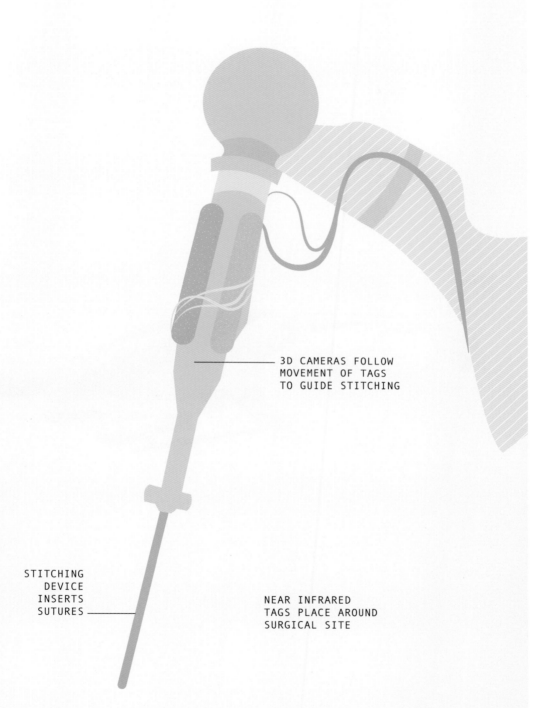

3D CAMERAS FOLLOW
MOVEMENT OF TAGS
TO GUIDE STITCHING

STITCHING
DEVICE
INSERTS
SUTURES

NEAR INFRARED
TAGS PLACE AROUND
SURGICAL SITE

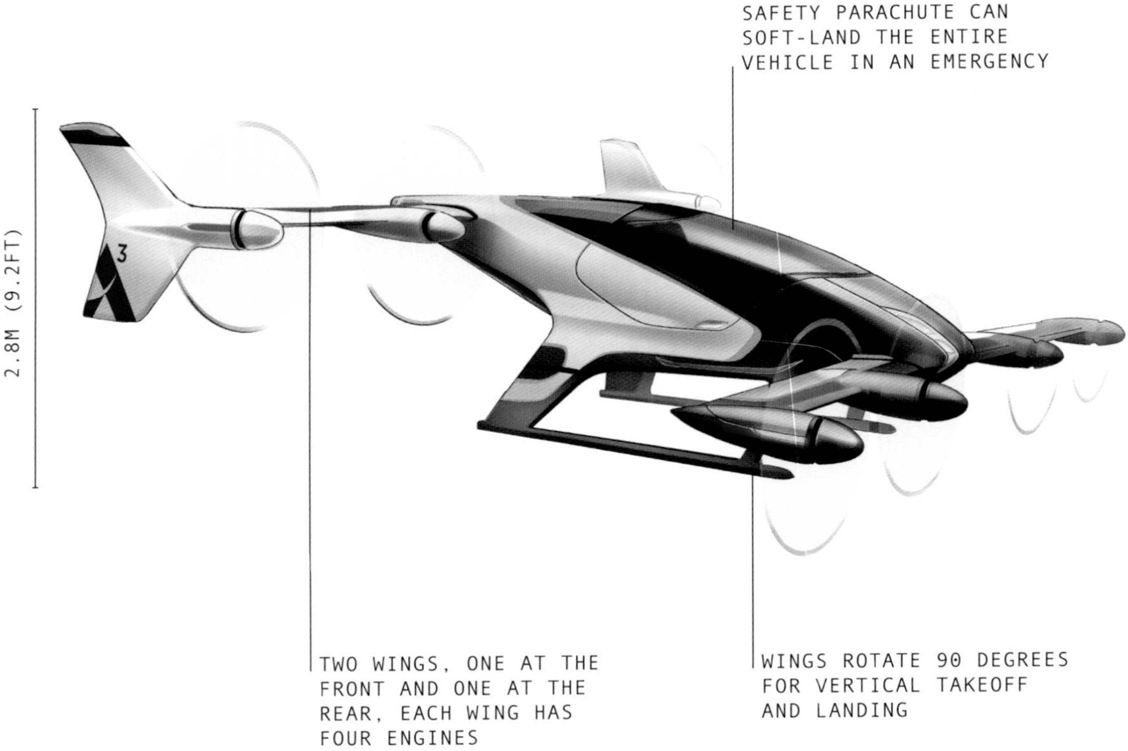

2.8M (9.2FT)

SAFETY PARACHUTE CAN
SOFT-LAND THE ENTIRE
VEHICLE IN AN EMERGENCY

TWO WINGS, ONE AT THE
FRONT AND ONE AT THE
REAR, EACH WING HAS
FOUR ENGINES

WINGS ROTATE 90 DEGREES
FOR VERTICAL TAKEOFF
AND LANDING

VAHANA

Height	2.8m (9.2ft) approx.
Weight	726kg (1,600lb).
Year	2018
Construction material	Composite
Main processor	Commercial processors
Power source	Battery

The Vahana is one of many planned 'flying cars' that have more in common with consumer drones than automobiles. Like a small drone, Vahana has several engines driven by electricity, and is self-piloted. It is not so much a flying car as an aerial robotic taxi. The name Vahana, incidentally, comes from a Sanskrit word meaning 'that which carries'.

Flying cars have been technically feasible, if expensive, for some time. Two basic problems have held them back. One is that few people have the necessary pilot's licence, the other is that air-traffic control could not cope with thousands of aircraft flying over cities. Taking humans out of the equation solves both problems: you do not need to be a pilot to use Vahana, and the machine will infallibly obey rules, stick to its flightpath and keep out of the way of other vehicles – just like the fleets of delivery drones that plan to share the same airspace.

While it might appear to be a flight of fantasy, A³ is part of Airbus, a company with decades of experience in the aviation market. Electric power makes Vahana cheap and reliable, but limits its range because batteries only provide about one-tenth of the energy that a typical gasoline-based system can. Vahana will be used for short hops rather than long-haul international flights, and that suits Airbus perfectly. The machine could offer a quick way for people to get from cities to airports and vice versa.

The initial Alpha version, currently undergoing flight testing, will carry one passenger. Range, speed and passenger capacity will increase with the Beta version: Alpha will fly up to 30 miles (48km) at 125mph (200kmh), Beta 65 miles (104km) at 145mph (233kmh). Safety is a high priority. In addition to other measures, Vahana will carry an emergency parachute so the entire aircraft can descend for a soft landing if problems arise.

Unlike the traditional idea of a flying car, Vahana will not take off from a driveway. Instead, the passenger will likely get a taxi – presumably a driverless one, such as Waymo (see pages 79–81) – to the nearest helipad, where scheduling software will have arranged a Vahana pickup.

Vahana has a tilting wing design. It takes off vertically like a helicopter, then the two wings, each with four engines, rotate through ninety degrees, after which Vahana flies horizontally, like an aeroplane. Vahana has a sophisticated sense-and-avoid system using radar, cameras and other sensors to avoid other air users. As with delivery drones, its introduction will need a change in existing regulations governing unmanned aircraft flying without human control.

Whether the vehicle is commercially viable will depend on price, which should be far lower than existing air services thanks to the cheap electric propulsion and the absence of a pilot. Vahana suggests it will work out to around £1–£2 per mile (1.6km), similar to a road taxi, but will be two to four times faster – and without the risk of delays caused by traffic or roadworks.

If this pricing is accurate, early adopters may not just be VIPs looking for a quick trip to the airport. Being able to fly in urban areas, Vahanas are likely to be popular for cheap commuting alternatives . More practically, Vahana would be an ideal air ambulance, with its low cost, quiet operation and ability to land on a smaller site than a traditional helicopter. Police may find it similarly useful.

There is no doubt that a flying taxi could be a great convenience, but only for as long as supply outstrips demand. In the coming decades, what if travellers find themselves standing in a long queue at the heliport because of a major conference, and face a one-hour wait for a five-minute journey?

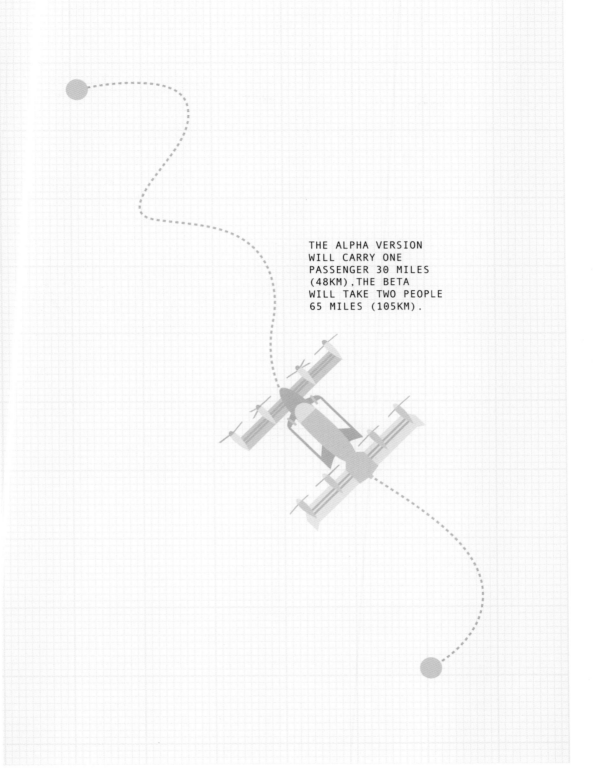

THE ALPHA VERSION
WILL CARRY ONE
PASSENGER 30 MILES
(48KM),THE BETA
WILL TAKE TWO PEOPLE
65 MILES (105KM).

ROBOTS
AT WAR

ROBOTS AT WAR

Perhaps the first military robots were eighteen-century fireships, unmanned vessels that crashed into enemy ships, bridges or fortifications to destroy them. The British called them 'machine-vessels' after the mechanical timer that was used to set off the fuse, and the Italians maccina infernale or 'machine from hell'.

Remote-controlled boats have been around since the 1890s, but navies have largely resisted using them. That may change with the Sea Hunter, an autonomous vessel designed to track submarines. Meanwhile robot submarines themselves are on the brink of a new era. Small, short-ranged, unmanned underwater vehicles have been around since the 1950s, but Echo Voyager is the prototype of a much larger vessel able to operate on its own for months on end.

The first military drones were arguably the balloons used to drop bombs on Venice during the Austrian siege in 1849. Franz von Uchatius, best known as the inventor of an early motion-picture projector, rigged small hot-air balloons to release bombs when a command signal was sent down a long copper wire. They were not successful; the technology was not up to the task.

Drones reappeared in the First World War, in the form of radio-controlled biplanes. These 'aerial torpedoes' carried an explosive charge to destroy enemy airships or ground positions, but neither the British Sopwith AT nor the American Kettering Bug was considered reliable enough to be used in action.

Drones have never been popular with air forces, but the Predator drone proved so effective in gathering intelligence during the conflicts in former Yugoslavia in the 1990s, that it was no longer possible to resist pressure from the intelligence community. The US Air Force reluctantly acquired a fleet of Predators. These started to carry their own missiles in 2001, after a series of incidents in which 'high value targets', including Osama bin Laden, were spotted, but air strikes could not be called in fast enough.

The Predator has since been superseded by the bigger and more powerful Predator B, better known as the Reaper. In military terms, it is a light reconnaissance aircraft with some weapons; the X-47B under development is a more serious combat aircraft. Meanwhile much smaller attack drones, like the portable Switchblade, are becoming important for tactical operations.

Remote-controlled ground vehicles have also existed since before the Second World War, but they have been largely limited to niche roles, such as bomb disposal, where machines like the PackBot 510 are essential. The Minitaur is a novel legged design offering greater mobility, which may prove valuable for urban reconnaissance. The XOS-2 is a more exotic development: a robotic exoskeleton giving its wearer enhanced strength and endurance. It will initially be used for logistics, but with the ultimate goal of a suit of powered armour that turns its wearer into a walking tank. The Covert Robot is equally remarkable, being a stealthy robot used to infiltrate enemy positions and move around silently without being spotted.

The only armed ground robots in service are South Korea's SGR-1A sentry robots. Russia experimented with remote-controlled 'teletanks' in the 1930s, and is pushing forward with a slew of armed ground robots, including Kalashnikov's 'Comrade in Arms'. The United States is more cautious, and armed vehicles like the Ripsaw are many years from service.

There are campaigns to actively prevent the development of 'killer robots,' not remote-controlled machines like the Reaper but autonomous robots able to find and attack targets without human oversight. The US military insists that they will always maintain a 'man in the loop', a human operator who needs to pull a trigger for weapons to be released.

Even if autonomous machines are banned, military robots are proliferating fast. Robots may replace soldiers on the front line, so the only humans involved may be commanders and drone operators on either side–and the civilians in the middle.

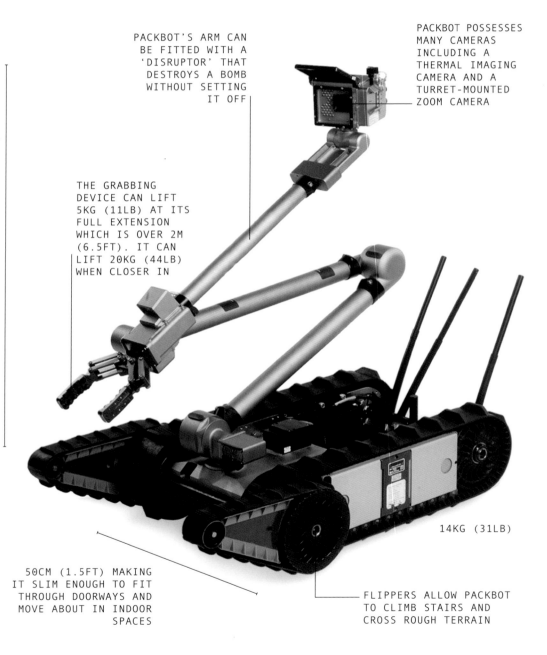

PACKBOT'S ARM CAN BE FITTED WITH A 'DISRUPTOR' THAT DESTROYS A BOMB WITHOUT SETTING IT OFF

PACKBOT POSSESSES MANY CAMERAS INCLUDING A THERMAL IMAGING CAMERA AND A TURRET-MOUNTED ZOOM CAMERA

THE GRABBING DEVICE CAN LIFT 5KG (11LB) AT ITS FULL EXTENSION WHICH IS OVER 2M (6.5FT). IT CAN LIFT 20KG (44LB) WHEN CLOSER IN

MAXIMUM EXTENSION IS OVER 20M (65.5FT)

14KG (31LB)

50CM (1.5FT) MAKING IT SLIM ENOUGH TO FIT THROUGH DOORWAYS AND MOVE ABOUT IN INDOOR SPACES

FLIPPERS ALLOW PACKBOT TO CLIMB STAIRS AND CROSS ROUGH TERRAIN

PACKBOT

Height	17.8cm (7in)
Weight	11kg (24lb)
Year	2002
Construction material	Steel
Main processor	Proprietary processors
Power source	Lithium ion battery

Bomb disposal may be one of the most dangerous jobs in the world. Any detonation is likely to be fatal, and improvised explosive devices (IEDs) can be incredibly tricky to disarm. Bomb-disposal robots save lives by widening the distance between the explosives and those trying to defuse them.

The first bomb-disposal robot was the 1972 Wheelbarrow, invented by Peter Miller, a former British Army officer. Designed to help deal with increasing numbers of car bombs in Northern Ireland at the time, Wheelbarrow was a simple remote-controlled tracked vehicle, based on a wheelbarrow with an electric motor. Its specific task was to attach a hook to a car bomb so it could then be towed away to a safer location. Since then, bomb-disposal robots have become increasingly sophisticated, gaining video cameras, robotic arms, disruptors and other accessories.

In 2002, the same year that they launched Roomba (see pages 63–5), iRobot® also launched a bomb-disposal machine – PackBot – which quickly proved its value in Afghanistan and Iraq. More than six thousand units have been delivered to date – sufficient to see the operation develop into a separate business, Endeavour Robotics®. Troops get quite attached to these mechanical heroes. One group of operators requested to have their 'fallen' comrade repaired rather than replaced with a new machine.

The current flagship model is the PackBot 510, but there are larger and smaller machines for different uses. Some can be tossed through the open window of buildings deemed too dangerous for humans to enter; others are massive machines for breaking into truck bombs. All are exceptionally rugged, to enable them to cope with the impacts and environmental extremes of military services. PackBot's developers discovered that this had to be built in from the start, and that it was impossible simply to strengthen an existing design. In order to reach a bomb, a robot may need to climb curbs and stairs and traverse rubble.

PackBot gets its name from the fact that it is packable; weighing just 11kg (24lb), it can be carried by one person and set up in two minutes. Mobile 'flippers' act as extensions to its tracks, enabling it to cross rough terrain and scale stairs, despite its small size. It can cross streams or ponds and even operate down to 1m (3.2ft) underwater. Being only 50cm (20in) wide, it fits easily through internal doorways.

PackBot is radio controlled and has sensors to give its operator the best possible view. These include multiple, high-resolution cameras for driving, a turret-mounted zoom camera, and thermal imaging. It has lights, both visible and infrared, and microphones.

Eight payload bays accommodate a variety of optional extras. These include a robotic arm known as a 'manipulator' with a camera and grippers to grab and move objects that at its full extension of over 1.8m (5.9ft) can lift 5kg (11lb). One of the commonest tools is a 'disruptor', a small, directional explosive device that destroys a bomb without setting it off.

The robot also has cable cutters. Real-life bomb disposal is never a matter of 'cut the red wire', but many bombs are detonated via a command cable. Cutting this is often the easiest way of removing the danger. PackBot can find and dig up buried objects with a probe fitted to its hull, as many IEDs are concealed under earth or sand.

The original PackBots were entirely remote-controlled and had no intelligence, but the machines are gradually growing smarter. The current version uses its arm to right itself if it overturns. If it loses radio communication, it will backtrack along its path automatically until its operator regains control. Future versions will be able to explore buildings on their own – a military version of domestic Roomba, if you like.

Police forces and firefighters also employ PackBots – not for suspected bombs, but for other potential hazards, involving chemicals or radioactive materials. One thing you can be sure of: every time PackBot goes into action, it is keeping a human well out of harm's way.

PACKBOT'S CAMERA HEAD
PACKBOT HAS MULTIPLE HIGH-RESOLUTION CAMERAS
THAT CAPTURE REAL-TIME IMAGERY AND VIDEO

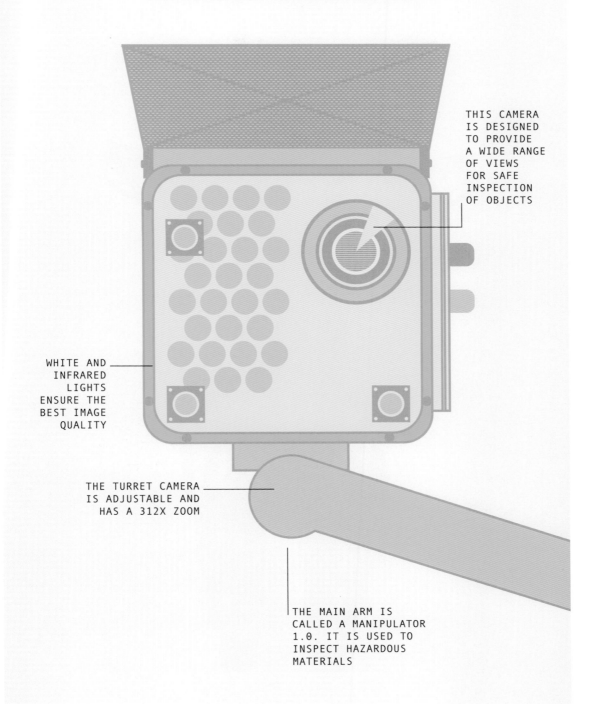

THIS CAMERA
IS DESIGNED
TO PROVIDE
A WIDE RANGE
OF VIEWS
FOR SAFE
INSPECTION
OF OBJECTS

WHITE AND
INFRARED
LIGHTS
ENSURE THE
BEST IMAGE
QUALITY

THE TURRET CAMERA
IS ADJUSTABLE AND
HAS A 312X ZOOM

THE MAIN ARM IS
CALLED A MANIPULATOR
1.0. IT IS USED TO
INSPECT HAZARDOUS
MATERIALS

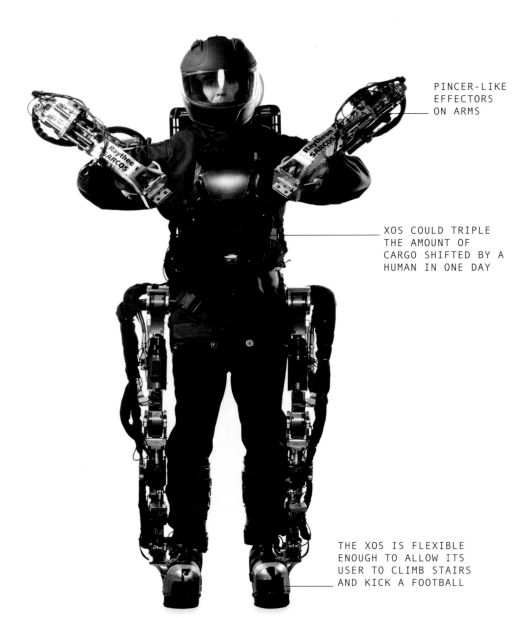

HEIGHT: DEPENDS ON WEARER

PINCER-LIKE
EFFECTORS
ON ARMS

XOS COULD TRIPLE
THE AMOUNT OF
CARGO SHIFTED BY A
HUMAN IN ONE DAY

THE XOS IS FLEXIBLE
ENOUGH TO ALLOW ITS
USER TO CLIMB STAIRS
AND KICK A FOOTBALL

95KG (209LB)

XOS-2 EXOSKELETON

Height	2m (6.6ft)
Weight	95kg (209lb)
Year	2010
Construction material	Steel
Main processor	Unspecified
Power source	External power

Ever since sci-fi 'dean' Robert Heinlein wrote *Starshi-p Troopers* in 1959, the powered exoskeleton has been a science-fiction dream for the military. Effectively it is a wearable robot in the form of the a suit. Memorably depicted as the 'power loader' that Sigourney Weaver's character used in the movie *Aliens*, Marvel's Iron Man also has a more advanced version. Stepping into an exoskeleton turns the mere mortal into a superhero able to lift heavy objects or smash through walls.

The XOS-2 (short for eXOSkeleton) developed by Sarcos Robotics™ is the real world's most advanced version revealed to date. The American developer unveiled XOS-2 in 2010, having developed an original model, XOS, in 2000 as the Wearable Energetically Autonomous Robot (WEAR). The XOS-2 was geared towards the military, specifically designed for the logistics mission – that is, loading and unloading cargo such as ammunition. Even today much of this work is carried out manually, and a soldier may shift as much as 7,000kg (15,430lb) per day by hand. XOS could triple this, effectively meaning that one soldier could do the work of three.

The XOS-2 weighs 95kg (209lb) and has elaborate hydraulics to magnify the operator's strength to lift weights of 90kg (198lb) with ease – the operator only needs to exert 1kg (2.2lb) of force to lift a 17kg (37lb) weight. Increased strength means greater endurance; the operator can shift heavy loads all day without getting tired. XOS-2 does not have hands, but mechanical pincers, known as 'effectors' for gripping and lifting objects. Because XOS-2 also amplifies leg strength, the operator can walk, run, or even stand on one leg, despite the weight of the suit. Multiple sensors in each joint make the XOS-2 responsive enough for climbing stairs and kicking a football, and the operator can recover their balance if they stumble.

XOS-2 is not fully mobile. The onboard power source is an internal combustion engine with fuel for half an hour, but it also needs an external hydraulic power source, and a flexible cable provides the high pressure for the powerful hydraulics. Most impressive is the intuitive behaviour. Operators get the hang of the exoskeleton within a few hours compared to the three to five days' training needed to drive a forklift truck.

After successful demonstrations in 2010, XOS-2 was shelved. Sarcos are currently working on a more sophisticated exoskeleton known as the Guardian XO. No specifications have been released, but a key difference is that the XO will be completely self-contained and run on battery power for six to eight hours. Ben Wolff, CEO of Sarcos, says that the dramatic improvement is not a matter of using better batteries, but making the suit more efficient. The XO consumes about one-tenth as much power as XOS-2, and this makes battery power feasible. In an echo of Leonardo da Vinci's work on humanoid robots, Wolff says that nature provided Sarcos with some of the most efficient solutions. The company has a history of making prosthetics, and biomechanics is one of its key areas of expertise.

The Guardian XO will launch in 2019. More compact, and operating in a smaller space than a forklift or crane, it should be a highly efficient load mover. Several XO operators will be able to work in a space normally occupied by a crane, and can move the exoskeletons around easily, as needed. The exoskeleton is also more intuitive and potentially safer than a forklift, and Wolff believes that it will be highly competitive in areas such as manufacturing, construction, logistics, warehousing and shipbuilding.

The Guardian XO will have no trouble meeting the military requirement for shifting ammunition, and the technology might even form the basis for actual powered armour, but will its launch come too late?. Combat robots are also developing rapidly, and by the time a real-life Iron Man suit is ready, it may already be obsolete.

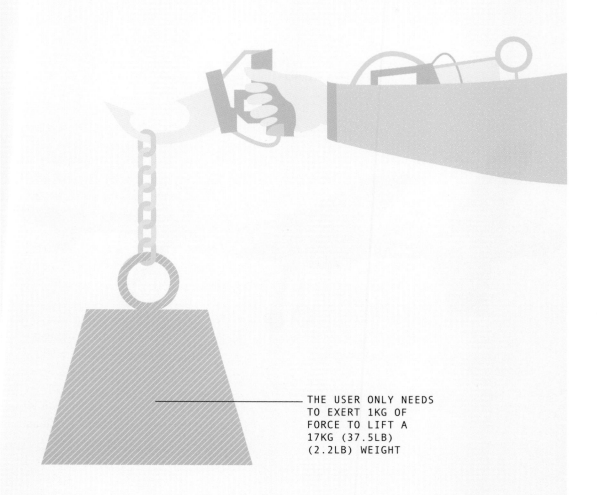

THE USER ONLY NEEDS
TO EXERT 1KG OF
FORCE TO LIFT A
17KG (37.5LB)
(2.2LB) WEIGHT

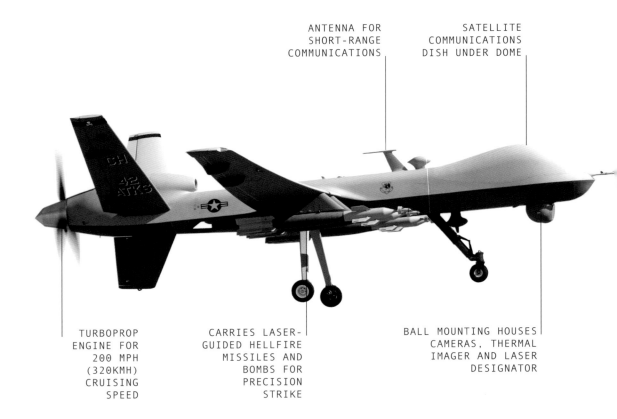

ANTENNA FOR
SHORT-RANGE
COMMUNICATIONS

SATELLITE
COMMUNICATIONS
DISH UNDER DOME

TURBOPROP
ENGINE FOR
200 MPH
(320KMH)
CRUISING
SPEED

CARRIES LASER-
GUIDED HELLFIRE
MISSILES AND
BOMBS FOR
PRECISION
STRIKE

BALL MOUNTING HOUSES
CAMERAS, THERMAL
IMAGER AND LASER
DESIGNATOR

11M (36FT)

MQ-9 REAPER®

Height	3.81m (12.5ft)
Weight	4.76mt (5.25t)
Year	2002
Construction material	Lightweight composites
Main processor	Proprietary processors
Power source	Honeywell TPE331-10 turboprop, 900HP (671kW)

The General Atomics MQ-9 Reaper is at the sharp end of drone warfare, the unmanned aircraft used by the United States and British military to carry out strikes against insurgents and terrorists across the globe.

In military aviation terms, the Reaper may seem an unimpressive aircraft. Its wingspan of 20m (65.6ft) is twice that of Lockheeds's F-35, the latest manned combat jet, but its takeoff weight is one-sixth as much. Reaper's top speed is 300mph (480kmh) compared to the F-35's 1,200mph (1,930kmh) and the drone has none of F-35's dogfighting agility. You will not see Reapers performing aerobatics at an air show, and they would be easy prey for modern radar-guided air defences. But there is one big advantage to the Reaper's light weight and broad wingspan: endurance. Like a glider, it can stay aloft with minimal effort, remaining airborne for twenty-seven hours at a stretch. It is an untiring eye in the sky for prolonged and detailed observation.

Developed from General Atomics earlier Predator, the Reaper is a remote-controlled aircraft rather than an autonomous one. It usually takes off from an air base a few hundred miles from the scene of the action and is guided by an operator on the ground. Once airborne, control passes over to a remote operating station, usually Creech Air Force Base in Nevada, which can control Reapers anywhere in the world thanks to its internal satellite communications dish. A two-person crew flies the Reaper: a pilot is responsible for flying the drone, while a payload operator handles the sensors and weapons. Shifts change every few hours, so several crews fly the same aircraft during any one mission.

A Reaper typically circles an area at 3,000m (10,000ft) – high enough that it cannot be heard and is unlikely to be spotted from the ground, but close enough for it to get a clear view.

The Reaper's main sensors are housed in a ball mounting, which rotates to any angle to keep a target always in view, whichever way the Reaper turns. This has thermal imaging and visible-light cameras with a 200x zoom on them, equivalent to a 12,000mm (39.4ft) telephoto lens on a 35mm camera. Some say it is able to read a licence plate from operating altitude. A Reaper may follow a vehicle or an individual, or observe a building or a village, carrying out a 'pattern of life' study and watching the comings and goings of various occupants. A Reaper may also carry radar and sensors to detect, locate and tap into radio signals below including mobile phones. Strikes may also be guided by radio trackers planted in a target's car or house by local agents.

The Reaper carries a mix of 500lb (225kg) bombs and Hellfire missiles, both laser guided. A laser designator in the sensor ball highlights the target, and the missile strikes where the laser is pointed. 'Super-ripple' firing means the Reaper's fourteen Hellfire missiles can be fired with just one-third of a second between missiles. The missiles are highly accurate, which raises two issues for the operators. The first is that they must be certain of their target; and the second is that Hellfire's 10kg (22lb) high-explosive warhead will kill or injure, not just the target, but anyone in the immediate vicinity.

Unmanned aircraft like the Reaper give intelligence agencies new options. They can enter areas in which it is too dangerous or too controversial to fly a manned plane, because there is no risk of losing a pilot and, if lost, drones can be denied. This means that intelligence agencies can track, locate and attack networks of terrorists in places that they could not otherwise reach, with little political risk.

Drones like the Reaper are not used for fighting conventional wars, but have opened up a new style of unconventional warfare. They have been highly effective at targeting terrorist leadership and have shown that there is nowhere to hide from drones. But is this enough to counter the fact that many people are uncomfortable over the civilian casualties and lack of accountability in drone warfare?

SHORT-RANGE ANTENNA ARE USED FOR
CONTROL BY THE LOCAL TEAM AT THE
AIR BASE FROM WHICH THE DRONE IS
OPERATING - ONCE IT IS AIRBORNE,
CONTROL IS PASSED TO THE REMOTE
TEAM IN NEVADA WHO USE SATELLITE
COMMUNICATIONS

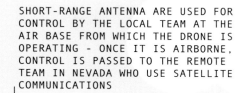

THE HELLFIRE MISSILE IS THE
SIGNATURE WEAPON OF A REAPER DRONE
STRIKE. IT HAS A RANGE OF 5 MILES
(8KM), AND ACCURATE ENOUGH TO HIT
WITHIN 1M (3.3FT) OF THE AIM POINT.
IT FLIES AT HIGH SUBSONIC SPEED,
AND VICTIMS USUALLY HAVE NO WARNING
BEFORE IT STRIKES

SENTRY CAN DETECT
'I SURRENDER' GESTURE

1.2M (3.9FT)

TRACKING DEVICE
AND LASER
RANGEFINDER

FIXED BASE MOUNT
- THIS VERSION
IS NOT MOBILE

SGR-A1 SENTRY

Height	1.2m (3.9ft)
Weight	117kg (258lb)
Year	2006
Construction material	Steel
Main processor	Classified
Power source	External mains electricity

'Spring-guns' have existed since the early days of firearms. At their simplest they are booby-traps that use a weapon, often a shotgun, with the trigger rigged to a tripwire, almost working – as they have been used – as a sentry soldier.

While the basic spring-gun was no more than a weapon attached to a wire, more sophisticated versions had several tripwires which can swivel to fire in the direction of the intruder. This brought them closer to being something like a robot. They became common in the eighteenth and nineteenth centuries, to prevent graverobbers from stealing bodies from graveyards. Known as cemetery guns they had the dual function of shooting at robbers and alerting the cemetery keeper to the intrusion. Cemetery guns were rented by the week by bereaved families to ensure that their loved ones remained undisturbed until their bodies were no longer of value to unscrupulous medical researchers.

The Hanwha Techwin SGR-A1, first installed in 2006, is a modern version of the spring-gun, a fully-automated robot sentinel to keep watch with unwavering concentration, twenty-four hours a day. It was built to help man the 160 mile (257km) 'demilitarized zone' (DMZ) between North and South Korea – one of the most heavily fortified borders in the world. South Korea has five thousand guard posts along this border, each with two guards, on shifts that change every two hours. That means an absolute minimum of twenty thousand soldiers are required to keep the border maintained. In practice, guard duty takes over forty thousand troops – about ten per cent of South Korea's entire combat force. Automating this duty will free up a huge amount of manpower.

The border already had plenty of sensors and surveillance systems, but the SRG-A1 is a gun turret able to engage targets and detect them. It is a stationary unit with three low-light-level cameras and a thermal imager, along with software for picking out moving objects – people and vehicles – and tracking them. The developers claim it can follow humans from a distance of 2.5 miles (4km) in daytime, or 1.25 miles (2km) at night.

The SGR-A1 challenges anyone entering the area, giving a person an opportunity to identify themselves and give a password or to put up their hands. The robot can respond to the correct password thanks to voice recognition, while gesture recognition can detect 'hands up'. If neither of these responses is detected, the robot can open fire.

The 'escalation of force' approach means the robot is likely to start with a nonlethal shot and then move up to lethal means if the target does not surrender. The robot is armed with a 5.56mm (0.22in) K4 machinegun and a 40mm (1.57in) grenade launcher. The latter can fire rubber bullets or other nonlethal 'kinetic' rounds. These only have a range of a few tens of metres, but if the target has been asked to give a password, it would presumably be close to the machine anyway.

There are plans to acquire one thousand SGR-A1s at a cost of a reported US$1 billion. A year-long pilot scheme in 2008 cast doubts over whether the system worked well enough, and a period of redevelopment followed, but by 2014 the system was apparently passed fit for service. An unknown number of units are now in operation, though details are limited for security reasons.

The SGR-A1 is controversial. Its manufacturers say it has a 'man-in-the-loop' mode, in which a human operator is responsible for pulling the trigger, as well as a 'man-on-the-loop' mode, where the operator just switches the robot on and lets it go. This second option means the SGR-A1 is perceived as a 'killer robot'.

The DMZ is peppered with landmines, which are also capable of killing without being directly told to by a human operator, but the SGR-A1 has attracted far more comment. This is perhaps because a stationary lethal autonomous system could evolve into a mobile one. Minefields stay where they are put, but fleets of robots locating and targeting humans on their own initiative is a potentially alarming prospect.

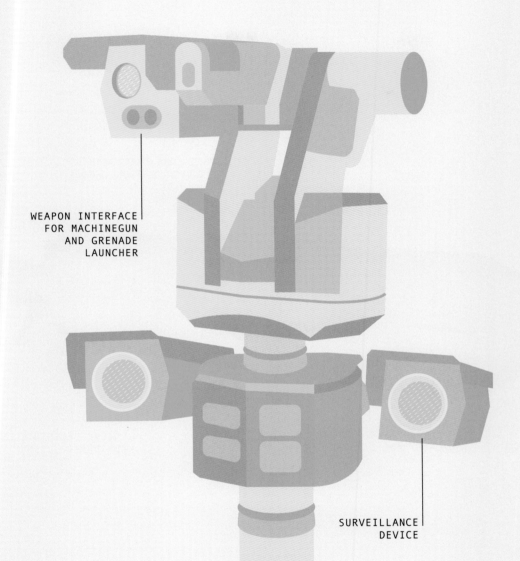

WEAPON INTERFACE
FOR MACHINEGUN
AND GRENADE
LAUNCHER

SURVEILLANCE
DEVICE

FOLDING PERISCOPE MAST WITH
SATELLITE COMMUNICATIONS AERIAL

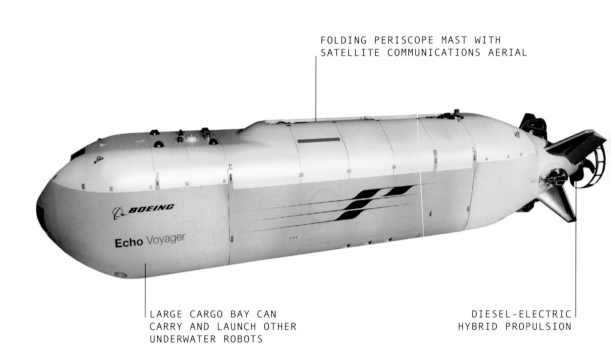

BOEING

Echo Voyager

LARGE CARGO BAY CAN
CARRY AND LAUNCH OTHER
UNDERWATER ROBOTS

DIESEL-ELECTRIC
HYBRID PROPULSION

15M (49.2FT)

ECHO VOYAGER

Length	15m (49.2ft)
Weight	31mt (34t)
Year	2016
Construction material	Steel
Main processor	Classified
Power source	Battery and diesel

It takes a special type of person to work on submarines, especially in wartime. Watery death is all too close, and the psychological effects of being in a confined, enclosed space for extended periods are considerable. Historically submarine crews have taken very high losses. It is a highly unnatural environment; so why not have the submarine run itself?

The unmanned submarine dates back to 1957 and the Self-Propelled Underwater Research Vehicle (SPURV) developed at the University of Washington. At 3m (9.8ft) long this machine could dive to 3,000m (9,840ft), far deeper than most manned submarines. Since then, this type of vehicle has been used by scientists and by commercial companies to inspect oil rigs and pipelines, as well as to help in wreck recovery and crash retrieval. Because only very-low-frequency (VLF) radio waves can penetrate seawater, unmanned underwater vehicles (UUVs) are controlled via a tether, or by sonar communication, which means they tend to stay close to their control ship.

Apart from a few exceptions, such as the underwater glider (see pages 199–201), UUVs are small vehicles with limited autonomy, and require a surface ship to launch and retrieve them. Boeing is seeking to break the mould with a 15m (49.2ft) unmanned submarine called Echo Voyager. It will carry out missions lasting six months with no human input, and no support vessel. Most of the operational cost is in the support ship, so a robot submarine operating on its own could radically change the industry. It can stay at sea and provide continuous reporting – on the spread of an oil slick, for example – without needing to return to a manned surface vessel to recharge every day.

Rather than just having batteries, Echo Voyager operates on a hybrid system of battery and diesel power: while underwater it uses battery power, but as it comes to the surface it switches to a diesel generator that recharges the batteries.

One of the ongoing challenges with Echo Voyager is autonomy. The robot must make its own decisions in demanding situations – for example, if it gets lost or when it detects another vessel on a collision course. 'The vehicle has got to understand what to do if it gets into trouble, make rule-based decisions and act in a way that allows it to stay safe and complete its mission', says Lance Towers, Director of Sea & Land at the Boeing Phantom Works.

The long endurance and range – over 7,000 miles (11,000km) – means that the Echo Voyager's onboard systems need to be far more reliable than those of other UUVs. Redundancy is built into critical systems: if one system breaks, a backup takes over so the Echo Voyager can make it back to port for repairs. Quite ingeniously, the Echo Voyager has a folding periscope mast that can be raised out of the water while the vessel remains submerged allowing it to communicate via satellite.

Echo Voyager's size is also one of its key features. Many times greater than other UUVs, it has room for an enormous payload bay that brings flexibility. UUVs are usually equipped for a single type of mission, such as surveying the seabed, but the Echo Voyager can carry tools and instruments for a range of tasks. It can also act as the mothership for a shoal of smaller unmanned submarines, releasing them underwater and retrieving them.

Echo Voyager was not designed for a specific mission, but the company hopes that, like previous projects, such as the 737 airliner, it will fulfil both civilian and military roles. There is potential here for such undersea robots to take on roles that include locating and neutralising underwater mines, launching unmanned aerial vehicles, tracking submarines, reconnaissance and what the US Navy refer to as 'payload deployment' which generally means dropping off sensors but can also include laying mines. In other roles, the Echo Voyager may find itself shedding new light on the dark deep of the sea, exploring and exploiting new mineral resources and examining ecosystems in unprecedented detail. Either way, its creation will almost certainly lead to advanced models, vessels that will not be tied to a ship but can set off on their own.

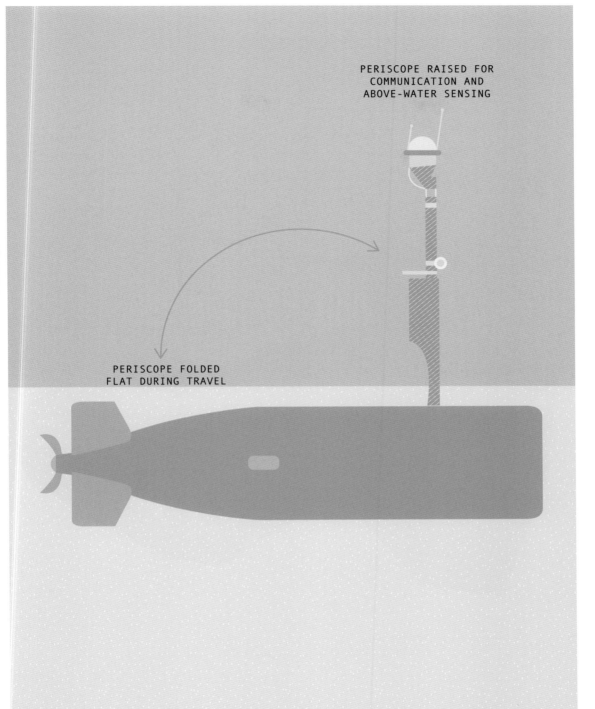

PERISCOPE RAISED FOR
COMMUNICATION AND
ABOVE-WATER SENSING

PERISCOPE FOLDED
FLAT DURING TRAVEL

MACHINEGUN REMOTELY
CONTROLLED BY HUMAN
OPERATOR

TRACKS FOR
MAXIMUM MOBILITY

LIGHTWEIGHT
CHASSIS

WINGMAN

Height	1.78m (5.8ft)
Weight	4mt (4.4t)
Year	2000
Construction material	Aluminium
Main processor	Proprietary processors
Power source	6.6 duramax diesel 750HP

In 2001, the US Air Force eventually accepted the idea of missile-firing drones – though only after the CIA threatened to develop its own – the US Army, however, has taken a more cautious approach. A project called Wingman aims to build confidence in armed unmanned ground vehicles, so they can stand in the front line in the place of soldiers.

Arming robots is easy enough. There were several robot weapon programmes in the 1980s, including the TMAP, a remote-controlled vehicle that could launch antitank missiles, keeping the operator away from enemy fire. However, in 1987, Congress prohibited spending on armed ground robots, believing that money was being wasted on over-optimistic projects.

Twenty years later, TALON SWORDS was deployed to Iraq. This was an armed version of a bomb-disposal robot, a bit like a giant PackBot (see pages 119–21) with a machinegun. It was a remote-controlled device and had no capacity to fire weapons on its own. It was never used, apparently for political rather than operational reasons, as there were concerns that the media would raise the alarm over 'killer robots'. US tactical armed robots have been stuck at the testing stage ever since.

The latest concept is a robotic 'wingman' for tank units. A new version of the M1 Abrams tank will feature an automatic loader. The crew member who previously loaded the main gun will now have a new role supervising robots. The Wingman will lead columns of tanks, making it the first to be hit by mines or ambushes. It will scout out suspected enemy positions, and will ultimately fire weapons under the direct control of a human operator.

The programme will build on previous developments, in particular the Ripsaw unmanned vehicle created by Howe & Howe Technologies Inc. This was built as a backyard garage project and submitted for a DARPA challenge in 2005 before being taken up by the US Army.

Ripsaw is the size of a sports utility vehicle (SUV) and weighs 4mt (4.4t), but has exceptional agility, accelerating from 0–50mph (0-80kmh) in 5.5 seconds, thanks to a six-litre diesel engine and a lightweight tubular chassis derived from NASCAR racing cars. This high speed allows Ripsaw to keep up with road convoys, and it can scale a 1.5m (4.9ft) obstacle or climb a forty-five-degree slope.

The US Army developed the sensors and software known as Safe Operations of Unmanned systems for Reconnaissance in Complex Environments programme (SOURCE) for machines like Ripsaw to drive around autonomously. SOURCE allows robot vehicles to drive at 30mph (48kmh) on roads or 6mph (9.6kmh) in busy urban terrain, avoiding vehicles, people and animals. SOURCE can even read and interpret road signs.

In the case of the Wingman, an operator will take control using a Head-Aimed Remote Viewer (HARV), resembling a virtual-reality headset which gives a robot's-eye view. When the operator turns their head, the robot camera rotates to follow.

The US Army has also developed an Advanced Remote/Robotic Armament System (ARAS) for its robots. This is based on existing turrets with some added features. If the weapon jams, it can be cleared remotely, for example. Rather than having one type of ammunition, there is a carousel of different bullet types that can be selected at will. The options may include standard rounds, reduced-range rounds for urban operations and 'nonlethal' riot-control rounds. ARAS can fire single shots with high precision; a fixed mount gives as much stability as a sniper rifle. It can also fire bursts or continuous fire as necessary.

The Wingman program envisages slow progress, and does not anticipate that a vehicle will be fielded until 2035. This will ensure a high level of confidence, and ensure that the robots are safe and remain under control. However, by that time battlefields could well be overrun by robots from other nations, such as the Russian Comrade in Arms, built by Kalashnikov (see pages 151–3).

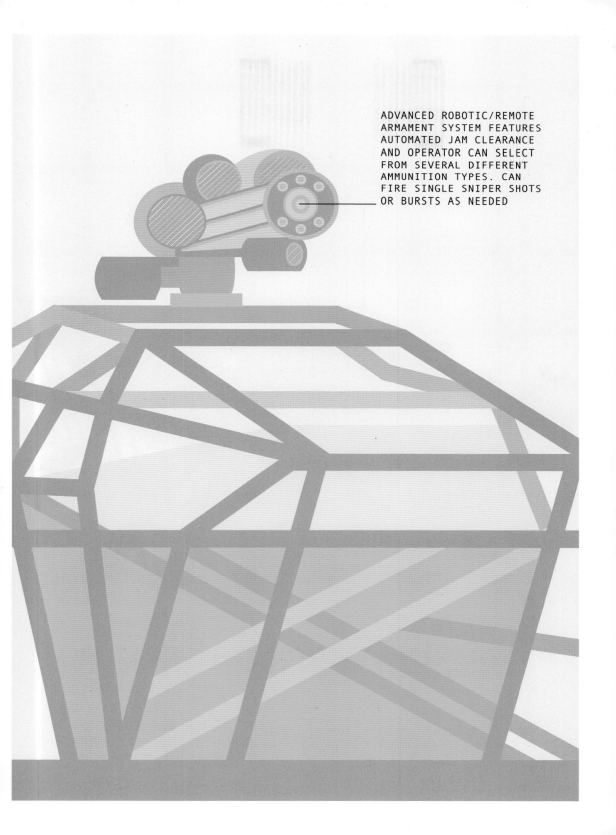

ADVANCED ROBOTIC/REMOTE
ARMAMENT SYSTEM FEATURES
AUTOMATED JAM CLEARANCE
AND OPERATOR CAN SELECT
FROM SEVERAL DIFFERENT
AMMUNITION TYPES. CAN
FIRE SINGLE SNIPER SHOTS
OR BURSTS AS NEEDED

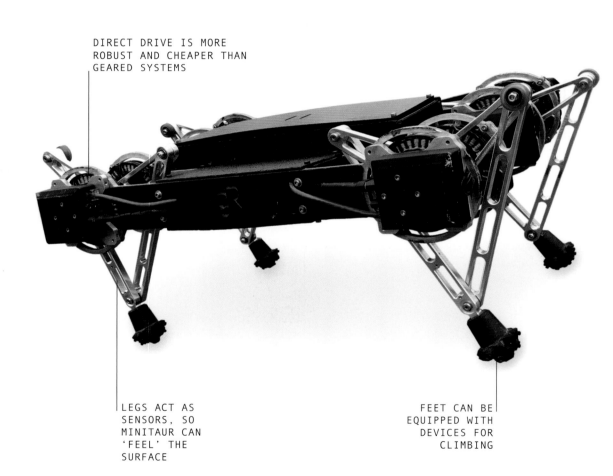

DIRECT DRIVE IS MORE
ROBUST AND CHEAPER THAN
GEARED SYSTEMS

LEGS ACT AS
SENSORS, SO
MINITAUR CAN
'FEEL' THE
SURFACE

FEET CAN BE
EQUIPPED WITH
DEVICES FOR
CLIMBING

GHOST MINITAUR™

Height	40cm (15.75in)
Weight	6kg (13lb)
Year	2015
Construction material	Steel
Main processor	NVDIA
Power source	Battery

Machines like PackBot (see pages 119–21) are fairly mobile, but they cannot compete with animals of a comparable size. Legs are the ultimate in ground mobility, and can cope with steep slopes, rugged terrain and obstacles that defeat even caterpillar tracks. Researchers have been trying to build mobile walking robots for more than a century, and their efforts are finally bearing fruit.

The BigDog robot developed by Boston Dynamics was the showpiece of the US Army's programme for a robot mule to carry soldiers' gear. This ran from 2005 to 2015; BigDog was described as 'the world's most advanced quadruped robot'. It looked great in demonstrations and racked up millions of YouTube views. Ultimately, though, BigDog lost out to wheeled designs for load carriers, as its innovative hydraulics were too noisy and expensive.

By contrast Minitaur is a dog-sized, affordable, practical robot. Its maximum speed, in a springy running mode on four legs, is about 5mph (8kmh). Not only can it cross rubble-strewn terrain impassable to wheeled and tracked robots, it can go up stairs and clamber over chain-link fences. Jiren Parikh, CEO of the makers, Ghost Robotics, says that it can even climb trees.

While robots like BigDog use hydraulics, Minitaur is entirely based on electric motors, which are simpler and cheaper. Its distinctive bouncy gait is closely linked to its main design features. The robot is powered by direct drive with no gearing. The company slogan is 'Robots That Feel the World™' and the motors themselves act as sensors – they are programmed to act like a spring-damper system, responding to changing forces. Even though the legs are actually rigid, the control system makes the robot bouncy.

The direct-drive system is agile enough to cope with slippery surfaces such as ice, or crumbling slopes. It responds almost instantly when a leg slips and makes the appropriate compensating movement to stabilise it. Minitaur can crawl on its belly, sidle sideways, or rear up against a vertical surface to reach objects above the ground, like an excitable dog at a kitchen counter. It can open doors and climb fences using manipulators on the ends of its legs for grasping. Minitaur's springiness allows it to jump up steps, or across a trench 50cm (20in) wide. Its advanced control system means it can even balance on two legs and walk like a biped for about twenty paces. In future, developers plan for it to learn new tricks, such as climbing different obstacles, with an artificial intelligence (AI) system.

The original version of the Minitaur was remote controlled, but the production version has stereo cameras and a high degree of autonomy. It can travel between points and avoid obstacles on its own. Batteries give enough power for more than four hours of operation, and the Minitaur can roam for about 10 miles (16km) on a single charge.

The first major customers are likely to be the US military, who have shown interest in Minitaur as a sensor platform for tasks that include bomb disposal and urban reconnaissance. Squads of Minitaurs could go ahead of troops and explore the interiors of buildings and tunnels. In the commercial world, the Minitaur might be useful in the mining industry for safety sensing, but agriculture may prove to be the biggest market. Farmers already use aerial drones for surveying their fields on a daily basis, to assess crop health. But drones are less useful for taking action, such as pulling up weeds, gathering soil samples or applying fertiliser. Gangs of Minitaurs could progress through fields carrying out routine tasks, recharging themselves from a base station as needed.

While Minitaur is still relatively untried, the design looks promising. The direct drive reduces the weight, cost and complexity. Having no gearbox means there is nothing to damage, making Minitaur highly robust. Plus, the electronics are all commercial off-the-shelf components, making it cheaper than similar robots. Minitaur may finally be able to deliver what legged robot researchers have been looking for over the past century: a practical, affordable machine, taking robots into uncharted territory.

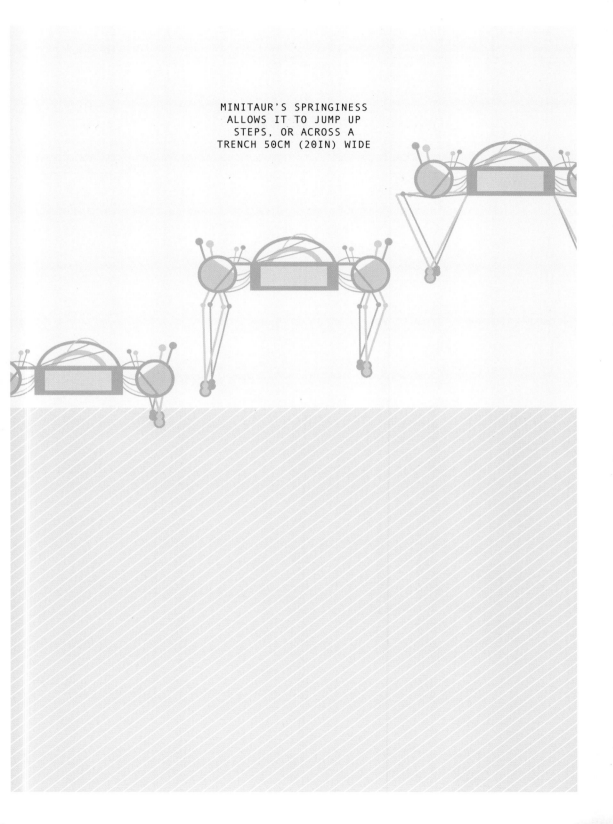

MINITAUR'S SPRINGINESS
ALLOWS IT TO JUMP UP
STEPS, OR ACROSS A
TRENCH 50CM (20IN) WIDE

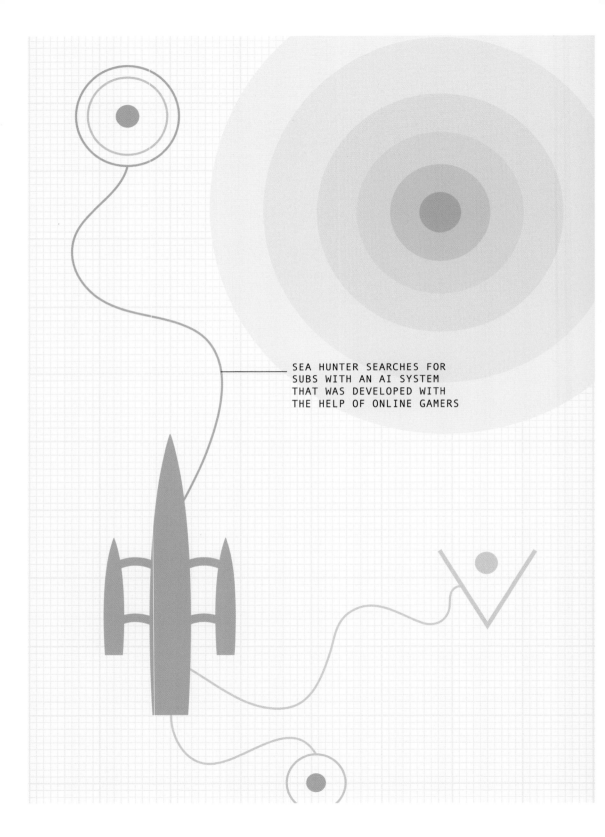

SEA HUNTER SEARCHES FOR
SUBS WITH AN AI SYSTEM
THAT WAS DEVELOPED WITH
THE HELP OF ONLINE GAMERS

SEA HUNTER

Length	40m (131ft)
Weight	135mt (149t)
Year	2014
Construction material	Composite
Main processor	Classified
Power source	Diesel

Submarines pose the biggest threat to warships in modern naval warfare. Armed with an array of missiles and torpedoes, they can sink anything without warning. While aircraft carriers may provide global striking power, they can only operate behind a protective screen of surface ships, aircraft and friendly submarines to keep them safe from underwater attackers.

The commonest vessels for antisubmarine warfare are destroyers. They bristle with active and passive sonar and other sensors, and are armed with torpedoes and missiles. A typical destroyer is 150m (500ft) long with a crew of three hundred, and of course they are expensive to acquire and operate. Navies can only afford a few destroyers, and they have a difficult job when looking for hidden opponents in the vast oceans.

Enter Sea Hunter. This submarine-hunting robot ship can do the job of a destroyer without the manpower, and at a fraction of the cost. It is equipped to locate the latest and quietest submarines, and has the speed and endurance to follow them wherever they go.

Sea Hunter is a trimaran that resembles a giant Polynesian war canoe, with a slim hull and two outriggers for stability. It is made of lightweight carbon-composite materials and was developed by DARPA under a programme called Antisubmarine Warfare Continuous Trail Unmanned Vessel. With no human crew to accommodate, everything is remarkably compact: there are no cabins, bunks, galleys or other occupied spaces. The only concession to humanity is a detachable pilot house that has been added for sea trials, during which the Sea Hunter remains technically under human command. Much of the vessel is taken up with fuel tankage for 40mt (44t) of diesel – enough to carry out a continuous three-month patrol. Sea Hunter can operate in rough seas up to about sea state 6, which means waves 6m (20ft) high and officially described as 'very rough'.

Satellite communication keeps Sea Hunter in touch with human operators, but ships are harder to control remotely than aircraft because of the volume of sea traffic. Sea Hunter looks after itself, with radar and electronic systems to spot other ships and avoid collisions, and a camera-based system with optical recognition software as a backup.

Like any other ship, Sea Hunter has to follow the Convention on the International Regulations for Preventing Collisions at Sea, the maritime version of the Highway Code. This dictates when a vessel should give way. One rule is that manoeuvres must be visible to the operator of the other vessel, so Sea Hunter is programmed to make large, sweeping course corrections rather than the subtle turns that a navigation computer would otherwise adopt.

Though Sea Hunter is unarmed, it tows an array of sensors for submarine detection. DARPA will not give details, but says that, owing to Sea Hunters 'unique characteristics', it will be able to employ 'non-conventional sensor technologies'. This might simply mean that, unlike a manned vessel, it can be completely silent, making sonar more effective for picking up the slightest sound. DARPA say it will 'achieve robust continuous track of the quietest submarine targets'. Sea Hunter is equipped with artificial intelligence (AI) to play the cat-and-mouse game of submarine hunting against a clever adversary without assistance. DARPA crowdsourced some of the algorithms, setting up an online submarine hunting game for players to try out innovative ideas.

Trials of Sea Hunter are just about complete and its developers can already see how the robot ship could take on many other tasks beyond submarine hunting. 'What we've realized over the course of the programme is that it's a truck', says programme manager Scott Littlefield. 'It's got lots of payload capacity for a variety of different missions.'

Inventors have been offering remote-controlled boats to navies since Nikola Tesla's radio-controlled demonstration in 1898. For the most part, they have been unsuccessful. Combining, as it does, long endurance, autonomy and low cost, the Sea Hunter might just be the vessel to turn the tide.

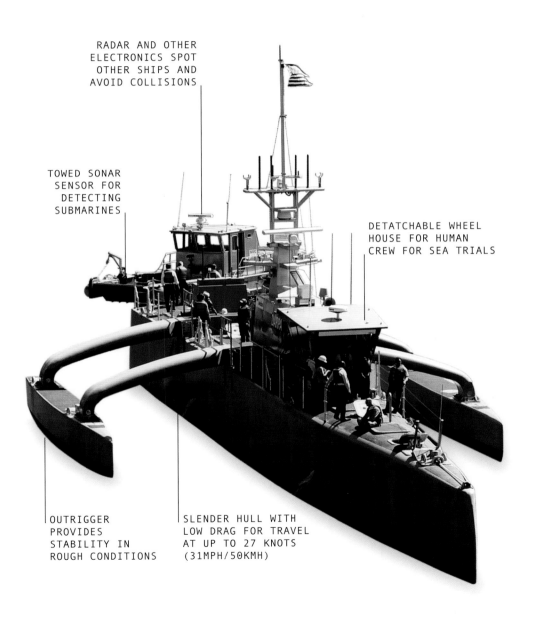

RADAR AND OTHER
ELECTRONICS SPOT
OTHER SHIPS AND
AVOID COLLISIONS

TOWED SONAR
SENSOR FOR
DETECTING
SUBMARINES

DETATCHABLE WHEEL
HOUSE FOR HUMAN
CREW FOR SEA TRIALS

OUTRIGGER
PROVIDES
STABILITY IN
ROUGH CONDITIONS

SLENDER HULL WITH
LOW DRAG FOR TRAVEL
AT UP TO 27 KNOTS
(31MPH/50KMH)

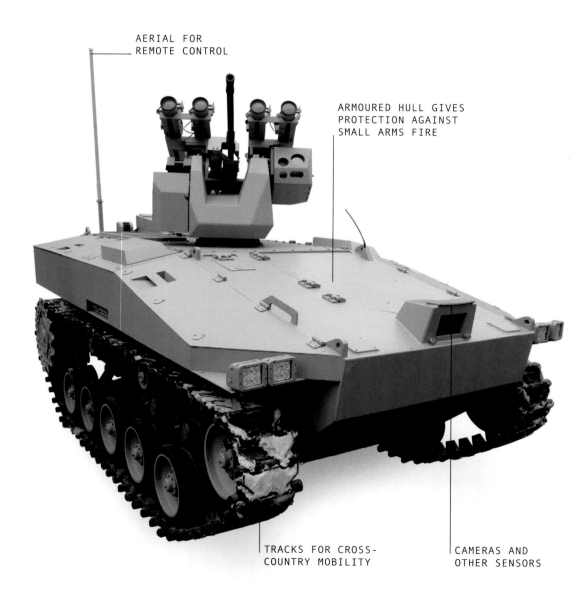

AERIAL FOR
REMOTE CONTROL

ARMOURED HULL GIVES
PROTECTION AGAINST
SMALL ARMS FIRE

TRACKS FOR CROSS-
COUNTRY MOBILITY

CAMERAS AND
OTHER SENSORS

COMRADE IN ARMS

Height	3.5m (11.5ft)
Weight	7mt (7.7t)
Year	2016
Construction material	Steel
Main processor	Classified
Power source	Diesel

There is no doubt that robots can swell the ranks of armed forces, which often find it difficult to get enough recruits. One obvious benefit from using robots is that they can take up positions in the front line where casualties are heaviest, reducing human losses. While many approach armed autonomous robots with caution, owing to the risks of 'friendly fire', Russia is taking a more robust stance. Its view is that such machines are not only inevitable, but that their deployment will be a good thing.

During the Cold War, Russia did not compete with NATO in sophisticated electronics. Instead, it churned out simple, but effective, weapons in massive quantities. The Kalashnikov assault rifle and the RPG-7 rocket launcher epitomised this approach. Russia had defeated a technically superior army during the Second World War by weight of numbers, but at the tragic cost of many lives. Perhaps it is this history that makes the Russian public sensitive to army casualties and the government keen to explore a robotic answer.

Russia was arguably the first to field armed robots, with the remote-controlled 'Teletanks' it used in the war against Finland in 1939. There had been little development since then until, in 2013, the industry was set in motion, with many companies pitching their combat robots to the Russian military. Some, such as the Stelok 'Sharpshooter', Platform-M and Boar, resemble bomb-disposal machines with guns, and are able to move around inside buildings. Others are the size of tanks and include the Whirlwind, which is a robot version of the BMP-3 combat vehicle, the Russian Army's main armoured personnel carrier.

The Soratnik or 'Comrade in Arms', developed by Kalashnikov Concern, was unveiled in 2016 and lies somewhere in the middle of the size range, being as big as a large car and weighing a hefty 7mt (7.7t). It is large enough for good mobility over rough terrain, but small enough to be procured in large numbers and issued widely. It runs on tracks and has a road speed of 25mph (40kmh). As its name suggests, the Comrade was designed to support infantry units in combat, and its turret is equipped with a machinegun – a Kalashnikov, of course.

It can also carry heavier weapons, such as grenade launchers. An alternative antitank version carries eight guided missiles. Being armoured, the Comrade is protected from small-arms fire and shell fragments. In fact, it is only vulnerable to large-calibre weapons and antitank missiles.

One defining facet is a secure communications system, which can be controlled from up to 6 miles (9.5km) away. It can carry and launch two small Zala drones, also made by Kalashnikov, to scout for targets. What really distinguishes it from Western equivalents is its software. You might expect a machine like this to be kept on a tight leash, but the Comrade has various levels of autonomy allowing it to take control of its own driving and weaponry. The Russian move towards autonomous robots comes partly from an acceptance that radio-frequency jamming is likely to be increasingly common on the battlefield. Communications are a weak spot for military robots, and many counter-drone systems work by jamming the link between the drone and the operator. If the link is lost, a robot needs to be able to act for itself. As far as the Russian are concerned that includes fighting solo.

At the extreme end, the Comrade can operate independently. Its manufacturers claim that it can detect, identify and engage targets – presumably telling friend from foe – from 1.5 miles (2.5km) away. With the ability to shut down into a 'sleep' mode to conserve power and wait for an opponent to appear for up to ten days, it allows a group of Comrades to hold a defensive position on their own for an extended period.

If Kalashnikov's ground robots prove to be as rugged and affordable as their assault rifles, they may start to fill the front ranks in future wars.

TURRET WITH MACHINEGUN
AND/OR MISSILES

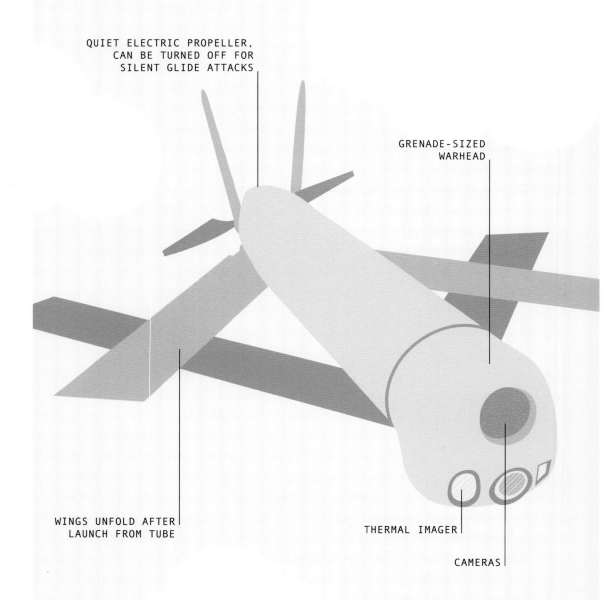

QUIET ELECTRIC PROPELLER,
CAN BE TURNED OFF FOR
SILENT GLIDE ATTACKS

GRENADE-SIZED
WARHEAD

WINGS UNFOLD AFTER
LAUNCH FROM TUBE

THERMAL IMAGER

CAMERAS

SWITCHBLADE

Wingspan	70cm (27.5in)
Weight	up to 2kg (4.4lb)
Year	2011
Construction material	Composite
Main processor	Classified
Power source	Lithium battery

The Switchblade drone is officially known as a 'tactical loitering munition'. It looks like a drone, it flies like a drone and it is controlled like a drone, using the same control unit. The difference is that Switchblade is designed strictly for one-way missions, diving into targets and destroying them with an explosive warhead. It is the first of a new breed of lethal robots that may change warfare – from boots on the ground to drones in the air.

Switchblade's origins go back to the early 2000s. Special Forces in Iraq and Afghanistan had started using small, hand-launched drones called Ravens, also developed by Californian company AeroVironment™. These were extremely popular because they provided a way of seeing over the next ridge or into the village ahead. Unlike the big Predator drones that were flown by the air force, Ravens were carried by foot soldiers and could be sent up at a minute's notice to check a perimeter or fly ahead of a convoy to spot ambushers.

The only problem with the Raven was that, while you could see the enemy, there was nothing you could do anything about it. Raven operators had to watch helplessly as insurgents set up mortars or took up firing positions to attack friendly forces.

Project Anubis, developed by the US Air Force for the US Army, aimed to develop a small drone. This resulted in AeroVironment's Switchblade, unveiled in 2011. The drone is carried in a tube like a miniature bazooka and launched into the air using compressed gas. After launch, its wings unfold – hence the Switchblade name. An electric propeller drives it for fifteen minutes at around 50mph (80kmh), giving a range of several miles. It sends back the stream from visible-light and thermal-imaging cameras via a secure, jam-proof data link. Once a target is located, the operator locks onto it and the Switchblade automatically pursues and destroys it, even if it is employing rapid evasion tactics.

Launching a Switchblade from behind cover, an operator can identify and engage a target several miles away, without ever being seen. The warhead is powerful enough to destroy lightweight vehicles, such as the pickup trucks favoured by insurgents, but is highly directional – it is known as the 'flying shotgun' – so it can hit its target, leaving people a few metres away unharmed. Switchblade can attack from any direction, including a vertical dive. This means most cover is useless, including trenches and foxholes. It can enter buildings through windows or open doors.

One of Switchblade's key features is its 'wave-off' function. If, as the drone approaches, the operator sees the target has been misidentified – for example, it turns out to be an unarmed civilian – they can cancel the attack. The drone breaks off and circles, ready to find a different target. No other weapon can do this, allowing Switchblade to be used in situations where the 'rules of engagement' prohibit all other weapons.

Several thousand Switchblades have been used and they have reportedly been highly effective in action, although few details have been released. Many aspects of the weapon are still classified.

Switchblade now has many competitors, both in the US and elsewhere, for obvious reasons. A squad with Switchblades, either attacking or defending, can see and attack enemies without ever being spotted themselves. Actions fought with rifles from a few hundred metres away may become as rare as bayonet charges.

PORTABLE PRECISION STRIKE FOR THE FOOT SOLDIER

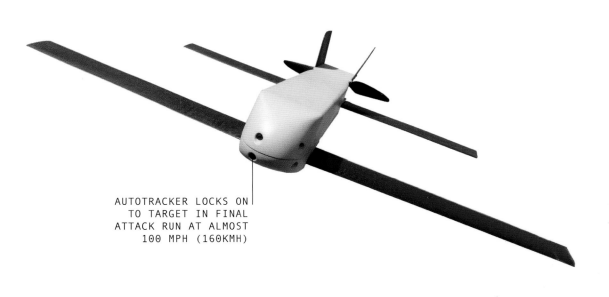

AUTOTRACKER LOCKS ON
TO TARGET IN FINAL
ATTACK RUN AT ALMOST
100 MPH (160KMH)

WINGSPAN 70CM (27.5IN)

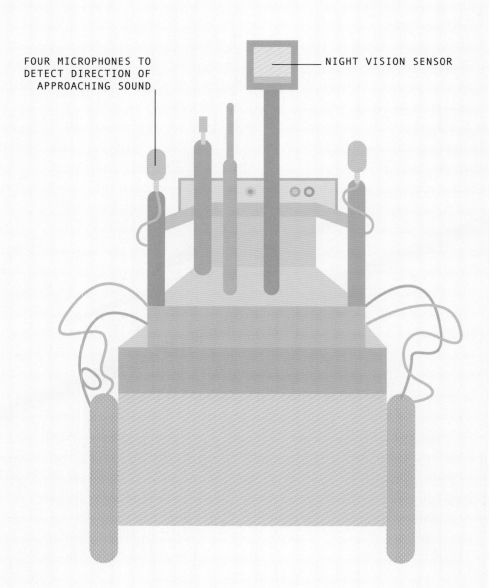

FOUR MICROPHONES TO
DETECT DIRECTION OF
APPROACHING SOUND

NIGHT VISION SENSOR

COVERT ROBOT

Height	60cm (24in) approx.
Weight	20kg (44lb) approx.
Year	2008
Construction material	Composite
Main processor	Commercial Processors
Power source	Lithium battery

Military ground robots are robust. By human standards, they are also conspicuous and clumsy. Sometimes, however, stealth is necessary – a requirement that led to the world's first robot spy.

In 2011 Lockheed Martin's Advanced Technology Laboratories at Cherry Hill, New Jersey, revealed details of what they called their Covert Robot, a machine able to evade detection by human sentries. It runs on four quiet rubber wheels, and expertly maps its environment by building up a 3D model of its surroundings, able to read lines of sight almost instantly. Its true secret weapon, though, is an ability to tell where it might be seen. It is in effect, a working spy.

The Covert Robot has four ways of avoiding detection. Most simply, at night it stays away from areas that are well lit and sticks to the shadows. Where there are known sentries, it avoids anywhere that is in their field of vision, utilising its four microphones to recognise approaching footsteps and determine their direction. When it hears someone coming, it moves out of their path, ensuring that it will not be seen, but also making sure that it has multiple routes of escape and avoiding being caught in a dead end.

The developers say that the robot has a 'multi-layered world model' – that is, it creates its own mental picture of the world, with obstacles, open paths, threats and light sources all factored in. The robot assesses every possible route for chances of being spotted, and weighs them up against its objective. The robot will take a risky route, but only if absolutely necessary, preferring to take lengthy diversions to reduce the risk. This world model changes second-by second, responding to new opportunities such as open doors, or new threats, including patrolling guards.

The first Covert Robot is a simple machine built at low cost out of commercial robot parts. It is essentially a test platform for trying out the software for covert behaviours in a real environment. A more advanced version will have more sophisticated hardware: it will be quieter and less conspicuous, as well as more mobile. This will at least include being able to climb stairs. The best hardware might be a humanoid robot like Boston Dynamics' Atlas (see pages 183–5), which could pass for human from a distance, making it much less likely to attract attention if seen.

The US military appears to have funded further work in this area, including development on ground robots for 'persistent surveillance'. These take up a position to observe a particular spot for a prolonged period, and can camouflage themselves – for example, using a robotic arm to cover themselves with leaves or branches to make them harder to see.

The US Navy has carried out more advanced work on deceptive software, aiming to emulate the deception and distraction techniques of animals. This includes behaviours such as small birds mobbing a larger predator even when they are too weak to harm it, and squirrels misleading others about where they have cached nuts by pretending to check non-existent caches. A Covert Robot might use similar techniques to lure guards away from their position, or to intimidate or bluff them into fleeing or surrendering.

These functions show an implied form of self-awareness: the robot has to consider how it appears to others. This human attribute, whilst vastly impressive, raises questions about the use of robots in warfare.

The Law of War requires soldiers to wear uniforms to distinguish them from civilians; anyone not identifiable may be an 'unlawful combatant'. There is no such law for robots, and a robot would not be troubled by the prospect of being captured and shot. The original fictional Terminator was an 'infiltration unit' able to imitate specific human voices. Real covert robots may have a whole range of tricks at their disposal to avoid and deceive humans to complete their mission. They may prove to be even more effective assassins than Reaper drones (see pages 127–9), but also far more ethically troubling.

DARK COLOUR FOR NIGHT
OPERATIONS

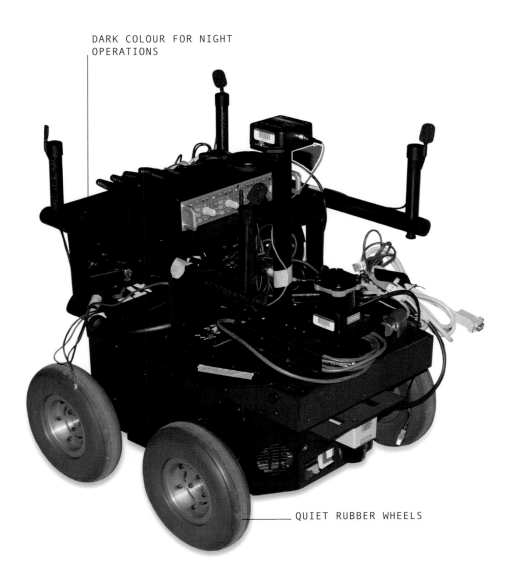

QUIET RUBBER WHEELS

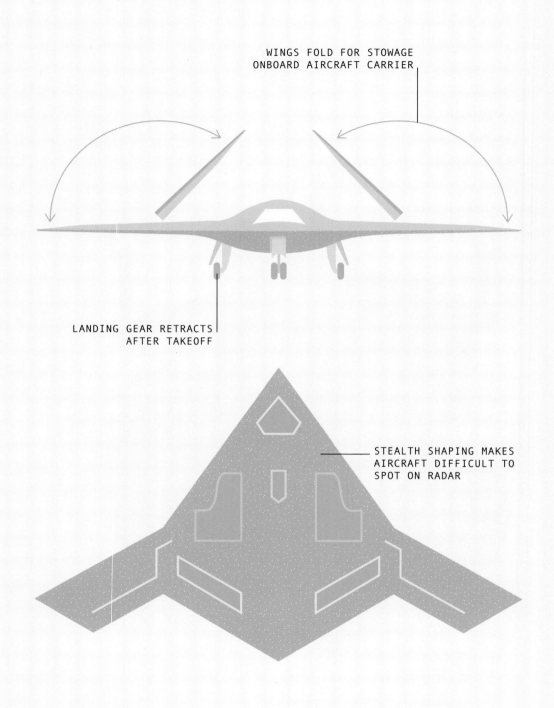

WINGS FOLD FOR STOWAGE
ONBOARD AIRCRAFT CARRIER

LANDING GEAR RETRACTS
AFTER TAKEOFF

STEALTH SHAPING MAKES
AIRCRAFT DIFFICULT TO
SPOT ON RADAR

GRUMMAN X-47B

Height	3.1m (10.1ft)
Weight	6.25mt (6.9lb), approx.
Year	2011
Construction material	Composite
Main processor	Commercial Processors
Power source	Pratt & Whitney F100-220U turbofan

In the opening stages of a conflict, one of the first priorities is to take out surface–to–air missile capacity so that attack aircraft can get through to hit targets. It is also one of the toughest challenges. This process of 'suppression of enemy air defences' is as dangerous as you might expect: planes have to fly right through missile defences ready and waiting to destroy aircraft.

It looks like a good job for a robot. Picking targets is relatively easy; radar sites and missile installations do not require any great discrimination by the pilot and there is little risk of hitting civilian buildings by mistake. Northrop Grumman's X-47B, a sleek jet that resembles a scaled-down strike aircraft with no cockpit, is an unmanned model designed for just this task. It has an angular profile and looks similar to manned F-22 and F-35 fighter jets; like them, it is fast and stealthy. It is difficult to spot on radar. First, its shape prevents radar waves being reflected to their source and, second, it is coated in special radar-absorbing materials. Such stealth does not make the aircraft invisible to radar, but allows it to slip past defences without being spotted until it is too late.

The X-47B tears through the sky at up to 12,000m (40,000ft) and travelling at 600mph (965kmh). It can find its target on its own, but a human operator provides mission instructions and clearance to release weapons. Those weapons comprise over 2,000kg (4,400lb) of smart bombs and missiles stored in an internal weapons bay – anything on the wings would spoil the stealthy curves.

The X-47B was developed for the US Navy, and must operate from an aircraft carrier, hence its foldable wings for easy storage. Flight tests were carried out in 2014–15 on the USS *Theodore Roosevelt*. Operations from the flight deck of a carrier are very different to those on an airfield: a steam catapult launches an aircraft into the air at takeoff; when landing, the plane must do so on a moving surface that is rocking from side to side. The high tempo of many carrier operations sees drones landing and taking off at the same time as manned aircraft, so they have to share airspace safely.

The X-47B passed its carrier trials with flying colours, but there are political problems. The US Navy's top priority, its F-35 and F-18 manned aircraft, gain all available funding. Anything that competes with them is likely to be in trouble. The programme has changed name and purpose several times since its inception in 2000. What started as an unmanned strike aircraft with some reconnaissance capability, became a reconnaissance aircraft with some strike capability that could also act as a tanker. In 2016 the requirement was further downgraded to a flying tanker with some secondary strike capability. For the time being, the drone will circle over a carrier and act as a flying petrol station for the F-35s and F-18s, which will continue to do the real work.

The latest change in role also saw the requirement for stealth being dropped. This will reduce costs, but also reduce the chances of the aircraft's use in combat missions against defended targets. And the production design will be something slower and cheaper than the original X-47B aircraft.

The US Navy argues that having a drone tanker will give them experience of working with unmanned aircraft on carriers, and that at some future time another generation of drones might play a more important role. When that will be is rather vague, and that is proof enough that Top Guns are not ready to give up their pilot seats to machines just yet.

CONCEALED EXHAUST TO
REDUCE INFRARED SIGNATURE

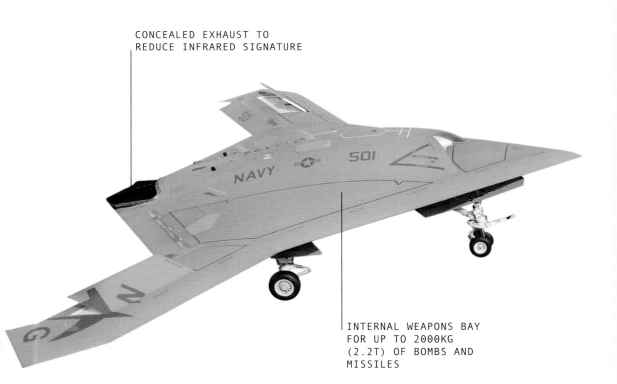

INTERNAL WEAPONS BAY
FOR UP TO 2000KG
(2.2T) OF BOMBS AND
MISSILES

12M (39FT)

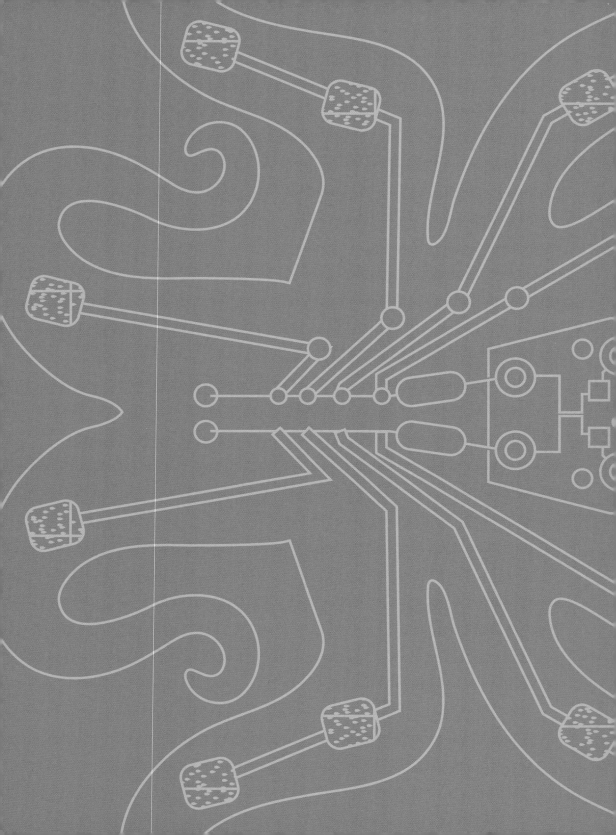

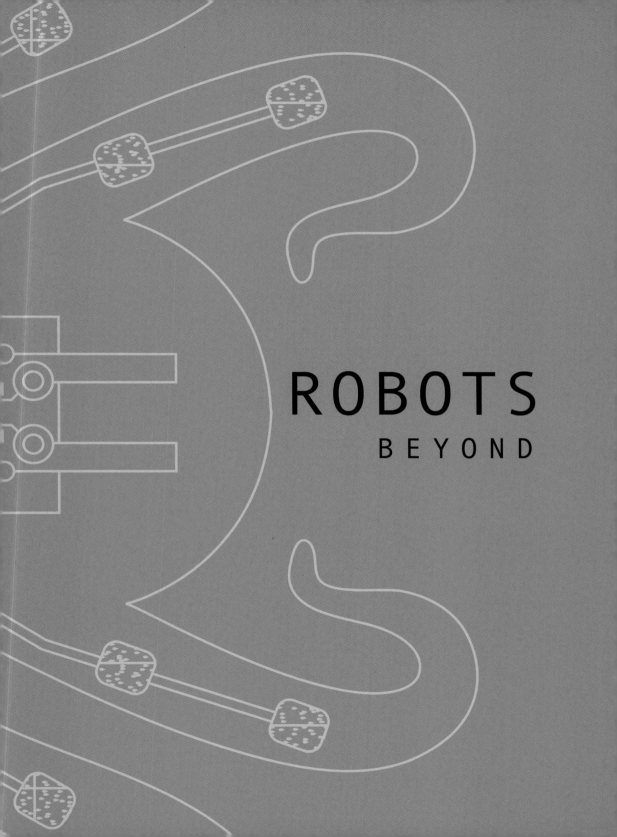

ROBOTS

BEYOND

ROBOTS BEYOND

While many of the robots already in service are awesome, they still fall a long way short of the robots dreamed of in science fiction – Commander Data, Terminator, the inhabitants of Westworld. But researchers are working hard to bridge that gap.

Robots do not need to look like humans, so most of them have functional shapes, like industrial robot arms. But when machines work in a human environment, opening doors and negotiating staircases, a humanoid form can make sense. Examples include Atlas, Boston Dynamics machine for human-type mobility, NASA's Robonaut for space missions alongside humans, and the US Navy's SAFFiR shipboard firefighting robot. While they are not quite Olympic athletes yet, they have come a long way in a few years. Robot bodies may not always be halting, inferior copies of human ones.

Robots are natural explorers. Unmanned systems were the first into space and machines such as NASA's Curiosity Rover are digging into the surface of Mars in search of life, decades before humans are likely to reach the Red Planet. The Flying Sea Glider can explore the oceans for months on end, while the Vishwa Extensor, a remote-controlled robotic hand for deep-sea divers, operates at greater depths than any human.

The human form is not the only one researchers want to copy. Dolphins can swim faster than submarines, and even leap clean out of the water. The best way to follow suit was to build a robot which mimicked nature's design – the amazing Dolphin, first of a new type of efficient swimming robots. The Ocotobot is a robot octopus, and although the resemblance is whimsical rather than practical, it highlights the technology of 'soft robotics'. Just as the boneless octopus is an alternative way of building a mobile, dextrous body rather than vertebrates like us, soft robotics explores the possibilities of machines with no rigid components.

Exploration is the story of sheer human persistence against the elements. A sea voyage lasting several months is an epic calling for tremendous endurance – unless you are a machine, in which case it is business as usual. Extremes of temperature, or even being underwater for prolonged periods, mean nothing to robots that are natural submariners. The Flying Sea Glider can fly for prolonged periods as well as swim. The Vishwa Extensor, designed as a remote-controlled robotic hand for human deep-sea divers, might end up equipping all sorts of unmanned craft.

Space is a better habitat for machines than humans. Unmanned systems were the first into space and the first to land on the Moon ahead of the astronauts. Machines such as NASA's Curiosity Rover are digging into the surface of Mars in search of life, and continue to work there decades before humans set foot on the red planet.

The Kilobot might seem like the least impressive machine of all, with less capability than a clockwork toy. It is in fact a research platform for testing software to control swarms of cooperative robots that can work together. This technique may ultimately dominate all forms of robotics, whether they are cleaning windows, performing surgery or working in fields or factories.

Nanobots are so far confined to science fiction: microscopic robots working in teams of millions to create or destroy. Nanobots in our bloodstream may one day detect and eradicate tumours before they can be a threat; others might tear down old buildings and reassemble them as new houses, efficiently recycling every particle of metal, stone and plastic. Technology has moved slower in this area than some expected, but that does not mean they are not coming.

This book gives some idea of what robots can already do, and what they will be able to do in the near future, based on what is already in the public domain. Government laboratories, and giant corporations like Google, may already have more advanced machines under wraps. As to where robots will go next, there are no fixed limits.

One thing we can say is that the technologies developed by individual programs – manipulators, mobility, swarming or social interaction software – will be combined to create more effective robots. Expect to see dolphin-like machines with hands like the Vishwa extensors, or Roombas with swarming software that allow several to cover a large house efficiently. Realistic but essentially static humanoids like Sophia may be matched with highly mobile bodies like Atlas to create a true android.

The other great unknown is artificial intelligence (AI). Ray Kurzweil, Google's director of engineering, anticipates that computers will exceed human intelligence in the 2040s. When that happens, current machines will look as primitive as da Vinci's mechanical knight.

Robots may be changing the world now, but they have barely started.

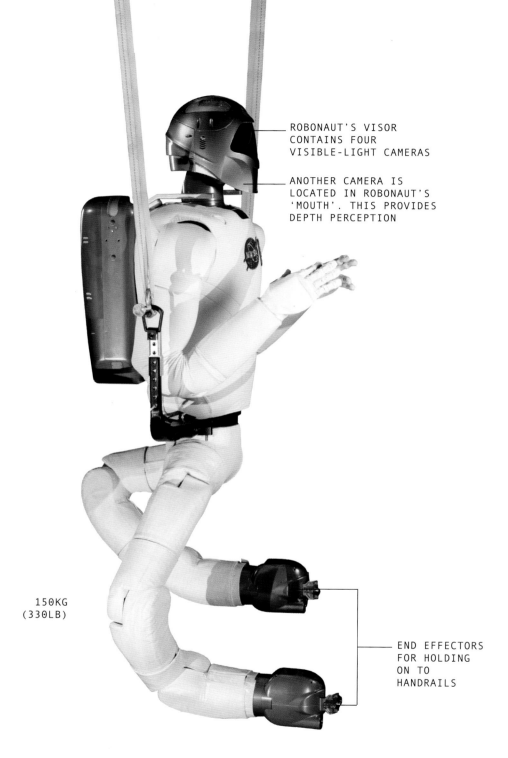

JUST OVER 1M (3.2FT) (WAIST TO TOP OF HEAD I.E NOT INCLUDING LEGS)

ROBONAUT'S VISOR CONTAINS FOUR VISIBLE-LIGHT CAMERAS

ANOTHER CAMERA IS LOCATED IN ROBONAUT'S 'MOUTH'. THIS PROVIDES DEPTH PERCEPTION

150KG (330LB)

END EFFECTORS FOR HOLDING ON TO HANDRAILS

ROBONAUT IS JUST AS DEXTEROUS AS HUMANS WITH 42 DEGREES OF FREEDOM IN TOTAL

ROBONAUT 2

Height	2.4m (7.9ft)
Weight	150kg (330lb)
Year	2011
Construction material	Composite
Main processor	PowerPC processors
Power source	Battery pack/external power

Robonaut is a robotic astronaut, not a replacement for humans in space, but a tireless helper. This autonomous, if unimaginative, assistant carries out simple, repetitive tasks so human astronauts can concentrate on more important things than housekeeping.

Extra-Vehicular Activity or EVA, which involves the astronaut donning a bulky space suit and leaving the International Space Station (ISS) to work on the outside, is demanding and exhausting. Astronauts only carry out EVA for a few hours at a time. It is also extremely hazardous, as the slightest tear in a spacesuit will suck all the air into the vacuum of space, leading to a perilous situation. Even if an astronaut is careful, micrometeors and orbiting debris are a constant threat. EVA is also impossible during solar storms due to high radiation levels.

By contrast, Robonaut does not mind vacuum. It is the same size and shape as a human, so it can use the full range of existing tools developed for astronauts, and is the ideal candidate for jobs that nobody else wants to do. The current version, Robonaut 2, was delivered to the ISS in 2011, although the Robonaut project dates back to 1997.

Robonaut's main talent is its dexterity, boasting the same skill at manipulation as a human. Its arms have seven degrees of freedom, while each five-fingered hand has twelve degrees of freedom and a gripping power of over 2kg (4.4lb). Even Robonaut's neck has three degrees of freedom, allowing it to rotate and tilt its head to look round, with four visible-light cameras and an infrared camera. Robonaut's backpack contains a power-conversion system that plugs into the ISS's electricity supply.

Robonaut's brain, which includes thirty-eight PowerPC processors, is located in its torso. It handles the input from over three hundred sensors around its body. Many of these sensors are in the hands: every joint has two position sensors, a pressure sensor to provide a sense of touch and four temperature sensors.

The eyes provide full 3D vision. At present, this only works well in good lighting conditions and struggles when conditions are less than ideal. The developers are seeking to improve it with the use of better object-recognition algorithms. While terrestrial robots work in unpredictable surroundings, everything in Robonaut's environment has been manufactured to a known size and shape. This should make it easier to catalogue and classify every object it needs to recognise.

Robonaut originally consisted of an upper body only, about 1m (3.2ft) tall, as what mattered was providing a pair of hands. It weighed 150kg (330lb), but getting around was not considered to be a problem in zero gravity. In 2014, a pair of legs, technically known as climbing manipulators, arrived at the ISS to make a Robonaut into a complete humanoid. Each leg has seven joints, and instead of having a foot it has an 'end effector' like a pincer, which can take hold of handrails like a climbing monkey. A vision system will be added to each of these, so Robonaut can see footholds to take hold of them.

The developers have experimented with other configurations, including a version of Robonaut for planetary exploration with a humanoid torso mounted on a wheeled chassis. This combination is known as Centaur.

For the present, Robonaut only takes on the most menial of tasks in the ISS: cleaning handrails and carrying out routine measurements of air circulation within the space station using a handheld meter. The robot also has its own 'taskboard', like a child's playset with buttons to push and switches to flip, as a practice device. Robonaut is not yet permitted to press real buttons.

Astronauts require a mass of life-support equipment to provide them with food, water and breathable air. A long-duration mission, such as a manned expedition to Mars, would rely heavily on robot labour to support the small human crew, so there would probably be more robonauts than astronauts. Unlike science-fiction stories, where robots invariably seem to rebel, they will probably be the most reliable members of the team.

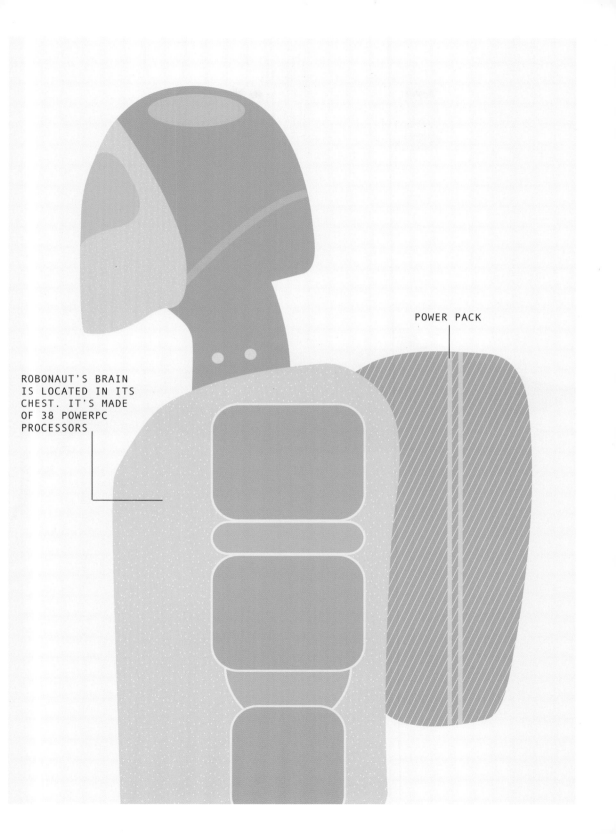

POWER PACK

ROBONAUT'S BRAIN
IS LOCATED IN ITS
CHEST. IT'S MADE
OF 38 POWERPC
PROCESSORS

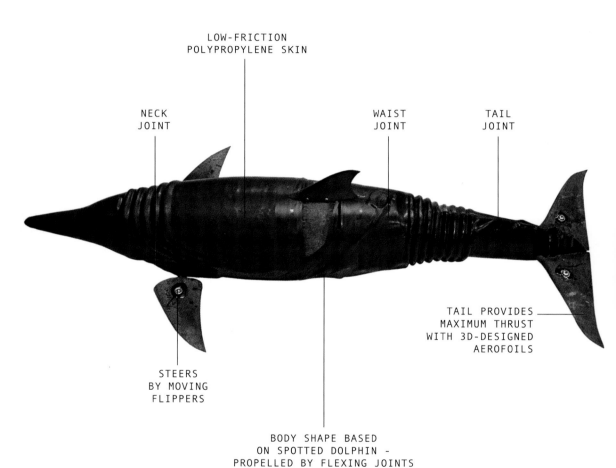

LOW-FRICTION
POLYPROPYLENE SKIN

NECK
JOINT

WAIST
JOINT

TAIL
JOINT

TAIL PROVIDES
MAXIMUM THRUST
WITH 3D-DESIGNED
AEROFOILS

STEERS
BY MOVING
FLIPPERS

BODY SHAPE BASED
ON SPOTTED DOLPHIN -
PROPELLED BY FLEXING JOINTS

72CM (28IN)

DOLPHIN

Length	72cm (28in)
Weight	4.7kg (10.3lb)
Year	2016
Construction material	Titanium and polypropylene
Main processor	ARM Cortex M3
Power source	Battery pack/external power

A Chinese underwater robot shaped like a dolphin, which moves by flexing its body and tail, is much faster than conventional rigid designs with propellers. It looks just like a real dolphin, and can even jump out of the water, pointing the way to future generations of agile, fish-like machines.

Water is a thousand times denser than air, and the drag it exerts makes submarines much slower than surface vessels. Boat designers improve speed by minimising contact between the hull and the water, and sailing yachts even have 'flying keels' with hydrofoils to lift them clear of the waves. This approach is useless underwater, so unmanned underwater vehicles are limited to a few knots. Reaching even that speed requires a disproportionate amount of power.

Nature is way ahead of us. Fish flick past effortlessly underwater, and dolphins can be seen 'porpoising', leaping out of the water in a series of high-speed leaps. In 1936, biologist James Gray had calculated that it ought to be impossible for dolphins to swim at 20mph (32kmh) because the underwater drag was too great. The puzzle was not solved until 2008 when another biologist – one Frank Fish – showed that the dolphin's tail is far more efficient at producing thrust than anyone had realised.

Roboticists have been borrowing ideas from nature ever since Leonardo da Vinci, and a team led by Professor Junzhi Yu at the Chinese Academy of Science in Beijing modelled its machine on a spotted dolphin. It is a scaled-down version, at 72cm (28in) long and weighing under 5kg (11lb). The design emphasises streamlining, with a tail designed for maximum thrust powered by electric motors. The dolphin's tail fins, known as flukes, and its flippers, are designed as 3D aerofoils.

The robot dolphin's body has three flexible, powered joints: a neck joint, a waist joint and a caudal joint between the body and the tail. These flex to propel the robot, while steering is provided by moving the small side fins. The parts of the dolphin's skeleton requiring strength are made of titanium, and the rest of it is aluminium and nylon, with polypropylene skin and flippers which allows the dexterity. A lithium ion battery provides power for three hours of operation.

The dolphin swims at an impressive 4.5mph (7.2kmh), or almost three body-lengths a second. From a design perspective, what matters is the robot's 'swimming number', the distance it travels with each stroke of its tail. In fact, it is close to the value for a real dolphin, a good indication that Yu's team has successfully reverse-engineered the dolphin's swimming technique. The robot dolphin is even able to copy the dolphin's feat of jumping clear of the water, something no underwater robot has ever achieved before. The advanced design required both a thorough understanding of dolphin aerodynamics, and the ability to translate them into a machine that could be built using existing materials.

The current robot dolphin is an early version, but can already run rings around propeller-driven underwater vehicles as easily as a playful dolphin circles a human swimmer. Yu's team is now looking at energy expenditure and the relationship between power and speed. This should lead to further increases in speed and efficiency, and longer, higher leaps. Maybe they will be the stars of a dolphinarium show soon!

More seriously, this type of propulsion should lead to underwater robots that swim as quickly and smoothly as fish, rather than churning the water inefficiently with propellers. They will be able to swim greater distances, manoeuvre more easily and manage bursts of speed that are impossible to current robots.

As well as carrying out scientific research, robot dolphins could take on industrial tasks, such as checking pipelines, locating fish and monitoring pollution. This form of propulsion is also far quieter than a propeller, so it is likely to be appealing to the military. Future submarines may find themselves being stalked by schools of stealthy, fish-like underwater robots.

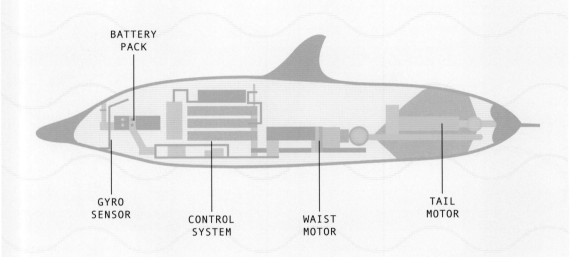

BATTERY
PACK

GYRO
SENSOR

CONTROL
SYSTEM

WAIST
MOTOR

TAIL
MOTOR

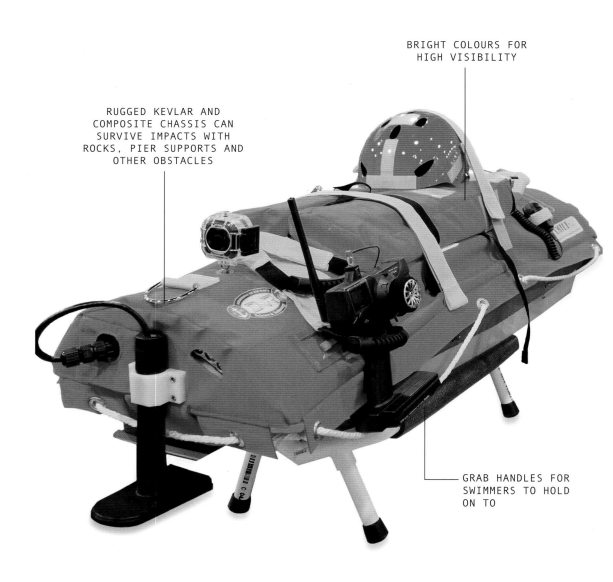

BRIGHT COLOURS FOR
HIGH VISIBILITY

RUGGED KEVLAR AND
COMPOSITE CHASSIS CAN
SURVIVE IMPACTS WITH
ROCKS, PIER SUPPORTS AND
OTHER OBSTACLES

GRAB HANDLES FOR
SWIMMERS TO HOLD
ON TO

EMILY

Length	1.2m (3.9ft)
Weight	11kg (24lb)
Year	2011
Construction material	Kevlar and composites
Main processor	Commercial processors
Power source	Battery pack

EMILY is a remote-controlled lifeguard that can swim faster than any human and that has already saved hundreds of lives. EMILY's name is short for Emergency Integrated Lifesaving Lanyard; it is 1.2m (3.9ft) long and weighs just over 11kg (24lb). Its size belies it power: EMILY has an engine like a jet ski, which sends the robot flying through the water at 22mph (35kmh), four times faster than the best Olympic swimmer. The design has no external propeller blades to tangle or cause injuries.

Made of Kevlar and aircraft-grade composite materials, EMILY is, according to inventor Bob Lautrup of marine robotics company Hydronalix Inc., 'virtually indestructible'. Its rugged construction means that it can survive almost anything; collisions with rocks or reefs at full speed cause no damage. Conditions that might pose a problem for a human lifeguard, such as 10m (33ft) waves, are no problem for EMILY.

The robot can be thrown off a pier or from a boat, or even dropped from a helicopter, then piloted by remote control to a person having trouble in the water. Being highly buoyant, with grab handles to hold on to, it can act as an emergency flotation device for up to six people as well as carrying additional life jackets in the case of large groups.

Adorned in a vivid orange, red and yellow colour scheme, and a flag that is visible above the waves, the robot also has lights for night rescues. A two-way radio allows rescuers to talk to people in the water – often necessary when they are panicking – as well as seeing them via a video camera. This is supplemented by a thermal imager, which helps to find people in the water at night or in severe weather. The robot can take a 700m (2,300ft) rescue line to a boat or person, and can even tow boats itself, a powerful attribute that helped rescue 300 Syrian refugees in 2015.

Lautrup says that EMILY was invented in 2009, after his team was testing a highly mobile waterborne robot at a beach in Malibu, California. They saw Los Angeles County California Fire Department lifeguards making rescues and realised that the robot would be ideal for sending flotation to a swimmer in trouble. The first version was fielded the next year.

There are already almost three hundred EMILY robots in service with coastguards and navies around the world. More sophisticated versions are on the way. One upgrade is a sonar sensor that can sense what is happening underneath the water, as the victim may not be easily visible from the surface. The challenge is to make the interface simple enough that the operator can see clearly what is happening in churning water. The team is also looking at other sensors to help rescue teams quickly locate someone who has fallen in the water from a ship or pier.

A project known as 'smartEMILY' is aiming to provide the robot with artificial intelligence (AI). This would help it to distinguish between people in the water who are conscious and active – and who would be able to hold onto the robot – and those who are unconscious and will need help from a human lifeguard.

Looking further forward, EMILY might be fitted with a soft manipulator arm or an inflatable device that can be placed under a swimmer who is unconscious or otherwise unable to hold on for themselves.

EMILY may be only one element of a team of robots. Some lifeguards are experimenting with using drones to patrol beaches and even to drop flotation devices. These could help locate swimmers and ensure that they stay afloat until EMILY can reach them and tow them to safety.

EMILY may not have quite the same appeal as the buff, bronzed members of the *Baywatch* team, and being rescued by a robot may seem a little unromantic. But when it is a matter or saving lives, EMILY is a star to match any human.

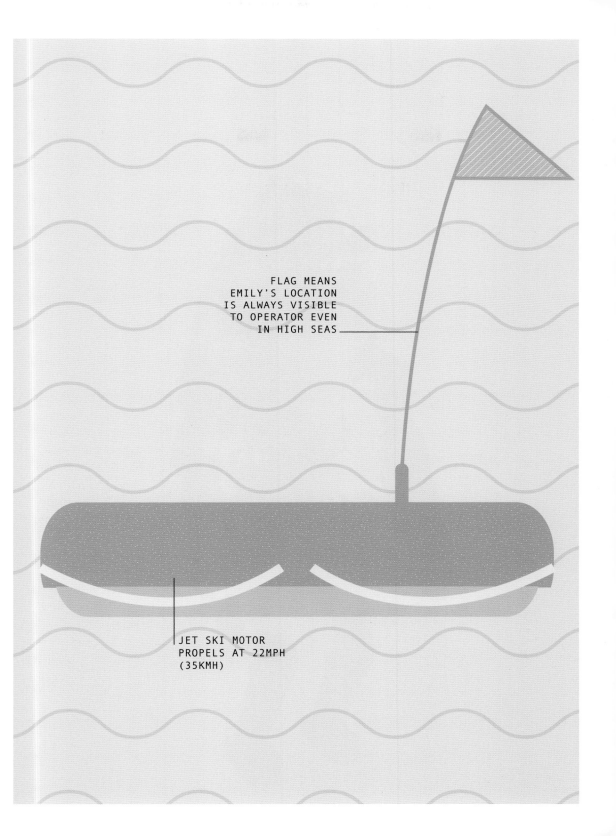

FLAG MEANS
EMILY'S LOCATION
IS ALWAYS VISIBLE
TO OPERATOR EVEN
IN HIGH SEAS

JET SKI MOTOR
PROPELS AT 22MPH
(35KMH)

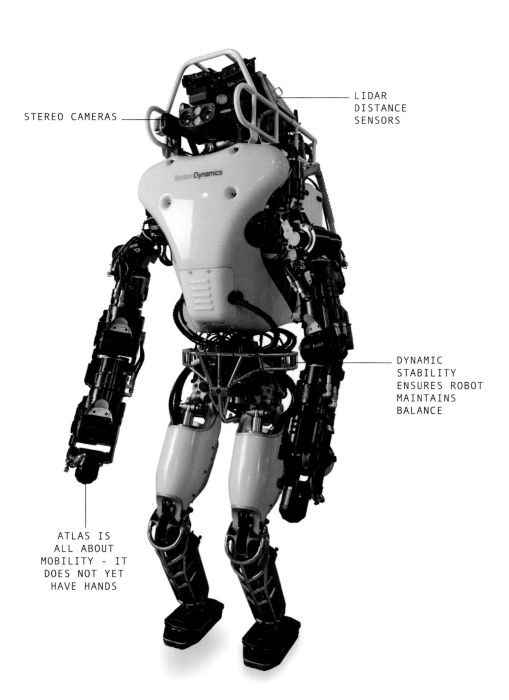

STEREO CAMERAS

LIDAR
DISTANCE
SENSORS

1.88M (6.12FT)

DYNAMIC
STABILITY
ENSURES ROBOT
MAINTAINS
BALANCE

ATLAS IS
ALL ABOUT
MOBILITY - IT
DOES NOT YET
HAVE HANDS

ATLAS

Height	1.88m (6.2ft)
Weight	75kg (165lb)
Year	2013
Construction material	Aluminium and titanium
Main processor	Commercial processors
Power source	Battery

Boston Dynamics is famous for machines that move and react to their surroundings in the same ways that living things do, and Atlas is their most advanced humanoid robot to date. Atlas is the heir to Leonardo da Vinci's robot knight, a step towards a general-purpose humanoid robot, with a level of agility that Leonardo could only dream of.

Many of the company's previous machines, such as the celebrated BigDog, have been quadrupeds. Staying upright is much easier on four legs than two. Atlas is humanoid, with twenty-eight joints, all in the same places as those of a human, driven by hydraulics and powered by electric motors.

Marc Raibert, the company's CEO, says that they do not directly copy nature, but rather they replicate what it does, an approach he refers to as 'biodynotics'. Most robots have static stability, staying upright by keeping their centre of mass directly over the support provided by their feet. But to move fast, you need to anticipate your forward motion and put your feet where they will be needed next.

'Think of the runner who wants to accelerate, putting their feet way behind them at the start of a race', says Raibert. 'Or the runner slowing down at the end of the race, putting their feet way in front of themselves, and leaning back. That is dynamic stability.'

Atlas has a somewhat peculiar gait, taking short, quick steps at regular intervals. It never seems to stand quite still. Even when it is not going anywhere it seems to march on the spot as it adjusts its footing. Jumping, running and hopping all rely on dynamic stability, as do the actions of dancers or gymnasts. Atlas has a good – though not perfect – sense of balance, and can recover itself and stay upright if it slips or is shoved. Like a human, Atlas adjusts its centre of gravity continuously and can cope with unsteady surfaces and uneven footing. If Atlas does fall over, it gets back to its feet without assistance. It can also pick up a heavy box and place it on a high shelf, a feat that requires a good mastery of balance.

Stereo vision and LIDAR range sensors (see page 44) allow Atlas to assess its environment and avoid obstacles. It requires a human operator, and Boston Dynamics are not specific about how much autonomy Atlas currently possesses.

Atlas can deal with the roughest of terrain, using its hands for support and balance where it needs to go all fours, to climb over or crawl under obstacles. The developers hope Atlas will ultimately be able to swing from one handhold to the next. Ultimately it should be able to climb rock faces or negotiate obstacle courses better than a human.

Atlas has not reached that stage yet, judging by the 2015 DARPA Robotics Challenge in which it competed. The challenge was inspired by the Fukushima disaster, and involved entering an area that might be contaminated by radiation. Robots had to operate controls that would stabilise a reactor, open and close valves and clear away debris. This required driving a vehicle, climbing a ladder, using a tool to break a concrete panel, and opening a door – all things that are easy for humans but difficult for robots. The results suggest that robots are not quite ready yet – for example, they had difficulty with some obstacles because they were unable to use handrails.

A new, more capable version of Atlas was unveiled in 2016 after the Grand Challenge. It is a work in progress; like many of Boston Dynamics machines, Atlas is a testbed rather than a finished product. Atlas itself may never be commercial, but there is no doubt that the technology – the dynamic stability, its agility, and ability to carry out human tasks – will feature in future robots.

'Our long-term goal is to make robots that have mobility, dexterity, perception and intelligence comparable to humans', said Raibert. Such machines could be emergency-response robots, versatile factory workers, household helpers . . . or almost anything else.

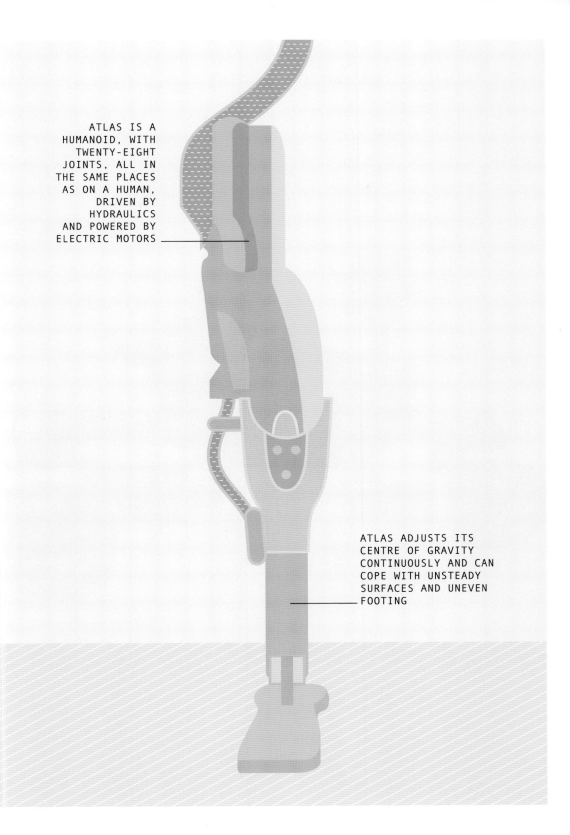

ATLAS IS A
HUMANOID, WITH
TWENTY-EIGHT
JOINTS, ALL IN
THE SAME PLACES
AS ON A HUMAN,
DRIVEN BY
HYDRAULICS
AND POWERED BY
ELECTRIC MOTORS

ATLAS ADJUSTS ITS
CENTRE OF GRAVITY
CONTINUOUSLY AND CAN
COPE WITH UNSTEADY
SURFACES AND UNEVEN
FOOTING

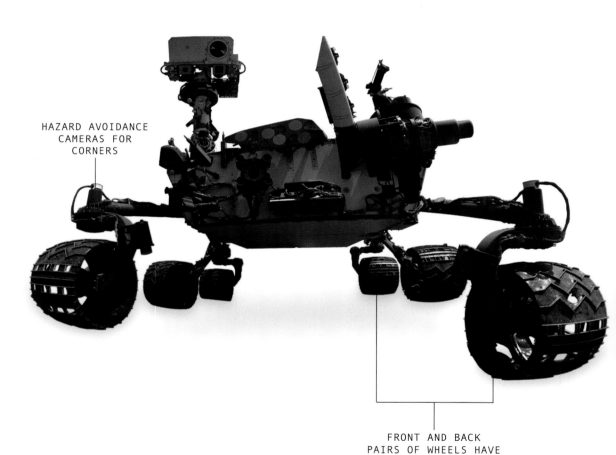

HAZARD AVOIDANCE
CAMERAS FOR
CORNERS

FRONT AND BACK
PAIRS OF WHEELS HAVE
INDEPENDENT POWER
AND STEERING

CURIOSITY

Height	2.2m (7.2ft)
Weight	899kg (1,981lb)
Year	2012
Construction material	Titanium
Main processor	BAE RAD750
Power source	Radioisotope

If robots are good for jobs that are 'dirty, dangerous and dull', then they are perfectly suited for missions to Mars. Dirty in the sense of continuous exposure to harmful radiation and dangerous because of the risks of space travel – twenty-six Martian missions have ended in failure – and dull, with the trip typically taking seven months. Robots also do not need to return to Earth, when a return leg would add tremendously to the cost and complexity of the mission. This is why the only Martian explorers to date have been robots.

Curiosity landed on Mars in August 2012 and is still going strong. It is the most advanced planetary exploration robot ever, with a wide array of scientific instruments to search for traces of life on Mars. It follows NASA's Sojourner in 1997, and Spirit and Opportunity in 2003. Sojourner weighed just 11kg (24lb), Spirit and Opportunity 180kg (396lb), but Curiosity is the size of a small car at 899kg (1,981lb) and 3m (9.8ft) long. It is also vastly more capable than its predecessors.

Curiosity has six large wheels, the front and back pairs having independent power and steering. Unusually, it is powered by radioisotope batteries. These would be too expensive and too environmentally hazardous for Earth, but they are good for Mars because they have effectively unlimited lifetime. Despite the nuclear power source, Curiosity moves agonisingly slowly: its top speed is about 4cm (1.5in) per second, less than half as fast as a tortoise.

Getting Curiosity to Mars took years of work and billions of dollars. One careless move might overturn the robot, get it stuck, or damage it, ending the mission. The distance between Earth and Mars means the time taken for radio signals to travel between them is between four and twenty-four minutes, so Curiosity cannot be controlled in real time. Each move is planned well in advance, and operations are conducted at a slow and careful pace.

Curiosity has no less than eleven sets of cameras, including a pair of hazard-avoidance cameras (hazcams) at each corner to aid automatic avoidance of unexpected obstacles. Two pairs of navigation cameras (navcams) on the mast give a longer view for route planning. There are various science cameras too, including a wide-angle panoramic camera.

Curiosity carries some 80kg (176lb) of scientific instruments. One device fires a tiny laser beam that can vaporise rock from 7m (23ft) away; amazingly the composition of the material it strikes can be determined from the spectrum of light emitted by the resulting puff of smoke. Curiosity's arm, which has shoulder, elbow, and wrist joints, can collect and examine mineral samples. The arm ends in a rotating turret with a set of tools. One is the Mars Hand Lens Imager, Curiosity's version of a magnifying glass, giving views of minerals detailed enough to distinguish features smaller than the thickness of a human hair. There is also a rock drill, a brush, and a device for scooping up samples of powdered rock and soil. Onboard tools include a neutron source to help detect water, believed to be an essential ingredient in life. There is even a miniature laboratory to analyse mineral specimens, and sample the Martian atmosphere.

Everything on Curiosity is designed for robustness, with a high degree of backup potential. For example, there are two central control computers. If one breaks down, the other takes over automatically.

Every discovery leads to new questions. Curiosity has literally only scratched the surface of Mars, drilling holes 5cm (2in) deep. Scientists now believe that they will have to go much further down to find any traces of ancient Martian life. More sophisticated rovers are already on the drawing board and in development.

Human astronauts will reach Mars one day. And when they do, it will have been explored, mapped, and analysed by robots like Curiosity.

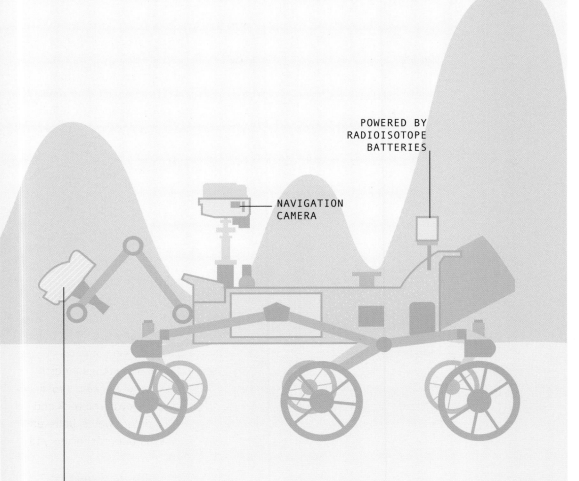

POWERED BY
RADIOISOTOPE
BATTERIES

NAVIGATION
CAMERA

TURRET AT THE
END OF ARM WITH
ROCK DRILL
AND SCIENTIFIC
INSTRUMENTS

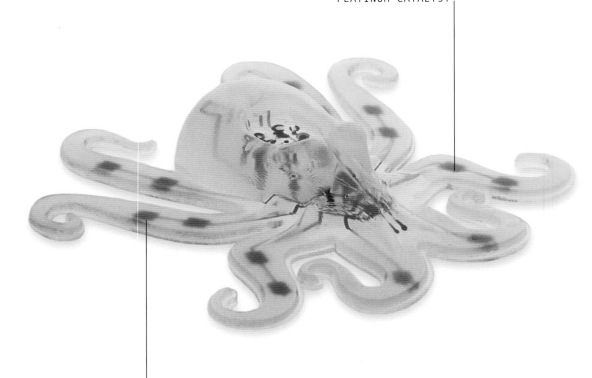

POWER PLANT DRIVEN BY
HYDROGEN PEROXIDE WITH
PLATINUM CATALYST

CHANNELS WITH
VALVES IN
TENTACLES

OCTOBOT

Height	2cm (0.8in) approx.
Weight	6g (0.2oz)
Year	2016
Construction material	Silicone rubber
Main processor	None (soft computer)
Power source	Chemical reaction

Robots, like other machines, tend to be constructed of hard materials, such as metal and plastic. The only flexibility is in their joints, which is why so much effort goes into how those joints work and how to control them. Not everything in nature is hard though, and there are plenty of soft manipulators, from octopus tentacles to elephant trunks, and even long tongues used to snag insects or tear up vegetation.

Soft manipulators can be much simpler and easier to control than their mechanical counterparts. Programming a robot hand to take hold of a doorknob is tricky, as the hand must be positioned in a certain way with all the fingers in the right places. A tentacle though, can simply wrap round the doorknob and envelop it. Soft robots are also better for handling delicate items, whether it's produce like fruit, or human patients in a hospital.

The most sophisticated soft robot to date is the Octobot, developed by Robert Wood and colleagues at Harvard University. At first glance it looks like a toy octopus, small enough to hold in your hand, and is made of 3D-printed silicone rubber components. Fluorescent dyes inside the translucent robot show up the tiny channels running down its limbs, making it look like a pop art sculpture. The Octobot does not just have soft limbs, but is an entirely soft machine with no electronic components. There are no batteries or motors or computer chips. Even the Octobot's 'brain' is a set of flexible microfluid circuits composed of pressure-activated valves and switches. Instead of electrons flowing around an electrical circuit, fluid runs around piping.

While previous soft robots have been tethered to a hydraulic or pneumatic tube, the Octobot is free roaming. It is powered chemically, via hydrogen peroxide reacting with a platinum catalyst producing water and oxygen. The resulting pressure inflates and extends the limbs, like water pressure inflating a rolled-up hose. Valves and switches inside Octobot's brain extend the arms in two alternating groups. When pressure rises at certain points, it closes one set of valves and opens another, directing flow into one half of the robot at a time. As the second side inflates it switches over again to the first side.

Octobot can carry out a preprogrammed series of actions, with up to eight minutes of power provided by one millilitre of hydrogen peroxide. It does not have any applications, but is intended as a demonstrator to inspire other researchers with the potential of soft robotics and show how challenges, such as self-contained propulsion can be overcome. It may also be the precursor to more sophisticated Octobots. 'Any future version would involve more complex behaviours, including the addition of sensors, and target simple modes of locomotion by introducing more articulation in the limbs', says project leader Rob Wood.

The other striking feature of Octobot is that its simplicity makes it very cheap to mass-produce: the components cost less than £2, most of which is for the platinum catalyst. Developer Michael Wehner says that this could open the door to applications that require large numbers of robots – cooperative search and rescue for example, with a swarm of soft robots squeezing through rubble to find survivors.

Part of the motivation for developing soft robots is that they are conformal: they can slip through narrow openings and shape themselves into whatever space is available. DARPA has looked at soft robots for infiltration – machines that might slither through letterboxes or under doors, or travel through air ducts. Tiny versions may have medical applications, as they can move more safely and easily inside a body than hard robots.

While a completely soft robot may not be practical, soft manipulators may be a viable alternative to hard robotic limbs. They will be safer to have working around humans than unyielding, rigid arms, and so may prove useful for applications with the elderly, for example. In a future where robots are likely to be rubbing shoulders more often with humans, the softer those shoulders, the better.

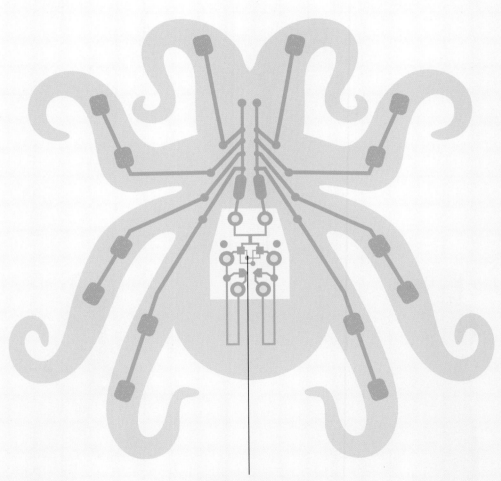

'BRAIN' MADE OF
MICROFLUIDIC CIRCUITS

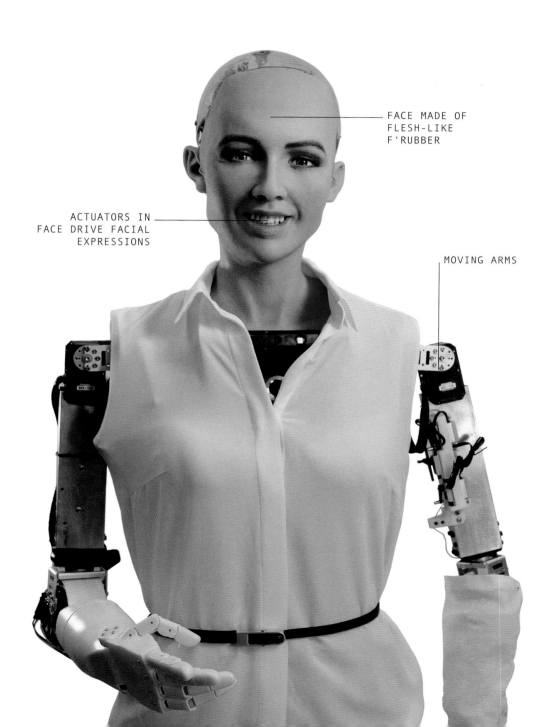

FACE MADE OF
FLESH-LIKE
F'RUBBER

ACTUATORS IN
FACE DRIVE FACIAL
EXPRESSIONS

MOVING ARMS

SOPHIA

Height	1.75m (5.75ft)
Weight	20kg (44lb)
Year	2015
Construction material	F'rubber face
Main processor	Commercial processors
Power source	Battery

There is something uncanny about Sophia. Her face is reminiscent of a movie star, specifically Audrey Hepburn, with – according to makers Hanson Robotics – 'high cheekbones, an intriguing smile, and deeply expressive eyes'. Sophia can convey emotion and has shown off her conversational skills on numerous chat shows. She is an interactive, entertaining robot.

Robots that imitate humans have a long pedigree, going back to Hero of Alexandria's automata in the first century. The art was further developed in medieval times with 'Jacks', mechanical figures that appeared to strike clocks or carry out other actions. Leonardo's robot knight is a particularly sophisticated version of this genre.

Walt Disney used the term 'animatronics' for the lifelike, moving machines he developed to populate his theme parks. One of Disney's first creations in the 1950s was an animatronic Abraham Lincoln, which delivered speeches prerecorded by an actor. The machine's lips moved in sync with the words, and it was programmed to produce facial expressions and gestures at appropriate points in the speech.

Sophia is the latest incarnation of the human-like robot. While the Geminoids (see pages 215–7) are remote controlled and have no brainpower of their own, Sophia has artificial intelligence (AI) – technology of the sort used by Internet chatbots – and aims to provide something like a natural human conversation with a machine.

Robert Hanson, CEO of Hanson Robotics, notes that large areas of the human brain are devoted to recognising and responding to facial expression. While the factoid that ninety per cent of communication is non-verbal is misleading, a robot that can communicate through facial expression, and that can understand another's facial expressions, should be an improvement on a simple voice from a screen. This may be especially true for children and the elderly who are not familiar with computers.

Appropriately enough, Hanson started out in Walt Disney's Imagineering company and has been perfecting the look of his robots since 2003, developing a special material that he calls F'rubber, along with a system of actuators to produce facial expressions and synchronise lip movements to speech. Hanson has not released details of Sophia's inner workings, but his previous robots had as many as twenty facial actuators. Sophia can move her head and neck and can gesture with her hands, which is all that is needed for conversation – and play rock, paper, scissors.

Sophia is essentially the hardware interface for a chatbot. Put simply, Chatbot is software that can understand and mimic conversation, and has proliferated the Internet since 2016. Apple's Siri and Microsoft's Cortana are both chatbot-based assistants that help millions of people on a daily basis. Chatbots are typically either based on a set of complex rules that require a major programming effort, or a learning approach that needs a huge database of conversational interactions. Like Siri and Cortana, Sophia's speech is machine-like and lacks the inflections of a human speaker.

Armed with scripted replies, added to her usual repertoire for each occasion, Sophia delighted, amazed and slightly appalled many TV hosts. ITV's *Good Morning Britain* presenters were clearly disconcerted in June 2017, with a rattled Piers Morgan commenting 'This is really freaking me out'. Host Jimmy Fallon made an almost identical comment on *The Tonight Show*, suggesting the uncanny valley effect is at work (see Geminoids).

Judging from Hanson's background, and the way that Sophia has been presented so far, show business may be her natural habitat. Disney's animatronic Abraham Lincoln was built to entertain more visitors per day than any actor could manage and robots like Sophia would be ideal as television presenters and interviewers within the industry. Thousands of organisations produce an streams of Internet corporate videos. A presenter who is wooden but flawlessly professional, and who can be rented along with the cameras, looks like competition for human hosts. Judging from the public reaction, Sophia is not ready for a wider public yet. That may well change in a few years.

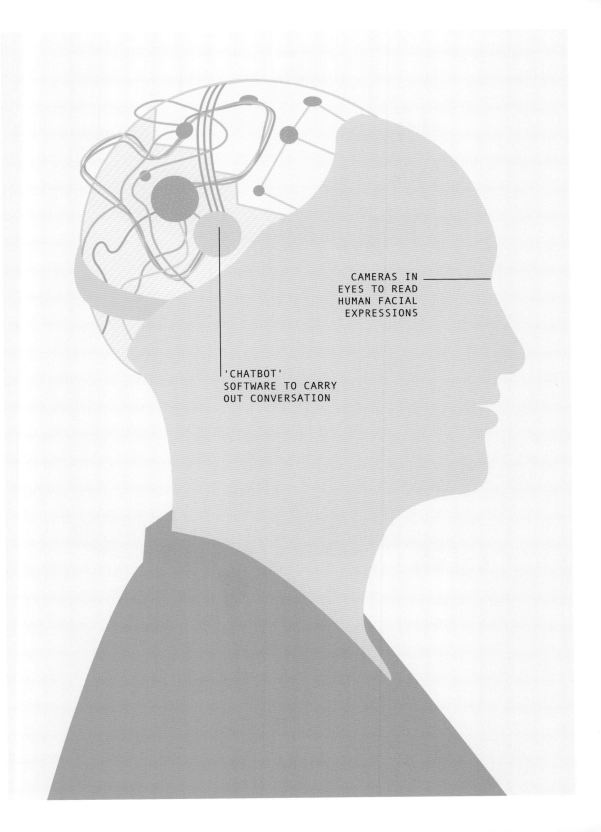

CAMERAS IN
EYES TO READ
HUMAN FACIAL
EXPRESSIONS

'CHATBOT'
SOFTWARE TO CARRY
OUT CONVERSATION

GPS AND RADIO
COMMUNICATIONS
ALLOW OPERATORS
TO TRACK THEIR
GLIDERS

INSTEAD OF
PROPELLERS
FOR UNDERWATER
TRAVEL, GLIDERS
HAVE A BUOYANCY
ENGINE

WINGS ENABLE
FLIGHT IN AIR
AND GLIDING
UNDERWATER

2M (6.5FT)

FLYING SEA GLIDER

Length	2m (6.5ft)
Weight	25kg (55lb)
Year	2017
Construction material	Composite
Main processor	Not specified
Power source	Battery

Underwater gliders are slow, steady unmanned submarines, carrying sensors to the furthest reaches of the oceans. They generally look like winged torpedoes about 2m (6.5ft) long. While most underwater robots are tethered and operated by remote control from a short distance away, gliders roam freely, surfacing occasionally to exchange data with an operator via satellite. They can travel on battery power for weeks or months at a time. A recent version can also fly, giving it speed as well as endurance.

Instead of propellers, gliders have a buoyancy engine. This pumps a small quantity of oil from an external bladder to an internal one, reducing the buoyancy so the glider starts to sink. Thanks to its wings, rather than simply falling, the robot glides forward several metres for every metre it descends, typically at less than 1mph (1.6kmh). It may drop to a depth of a 1,000m (3280ft) before pumping oil back into the external bladder, making the glider buoyant enough to start rising, this time gliding upwards at the same slow rate. It's a slow but frugal means of travel. In 2009, the Scarlet Knight glider operated by Rutgers University completed an Atlantic crossing on one battery charge, taking seven months.

Gliders are tough, sometimes returning from missions with multiple shark bites. Storms and high seas mean nothing to them. Fishing nets are the biggest hazard, but GPS and radio communications allow operators to track their gliders and they can often be reclaimed after some negotiation.

To date, gliders have watched underwater erupting volcanoes, inspected icebergs at close range and survived hurricanes. They tracked the Deepwater Horizon oil spill over an extended period, and carried radiation sensors to spot leakage from the Fukushima nuclear reactors. Gliders with acoustic sensors pinpoint tagged fish or listen in to whale song. Others carry out climate change research, measuring temperatures in different layers of seawater and mapping the abundance of algae. The Storm Glider lurks in hurricane-prone areas, bobbing up to take readings during extreme weather, while the Chinese make a glider that can explore the depths at 6,000m (19,685ft).

Gliders can be traced back to Doug Webb and Henry Stommel of the Woods Hole Oceanographic Institution. Since their beginnings in 1991, gliders have matured. In 2002 there were about thirty gliders worldwide. Now there are several hundred, with plans to build thousands more.

Their low speed is a limitation, but the Flying Sea Glider developed by the US Naval Research Laboratory could overcome that. As the name suggests, this is a glider that flies above the water as well as gliding through it. The same sort of aerodynamics applies in air and water, so it was a matter of finding a design that combined both into one airframe.

The Flying Sea Glider can be released from an aircraft or launched from a ship. A propeller driven by an electric motor can carry it to an operational area over 100 miles (160km) away. When it reaches the desired location, it dives vertically into the water, like a pelican diving for fish. Hollow wings and other spaces in the glider fill with water through a series of small holes, and it soon reaches neutral buoyancy. It can then proceed like any other underwater glider.

Flying gliders are ideal where several gliders need to be in place rapidly, for example to monitor an oil spill, a radiation leak or the path of a hurricane. The gliders could take measurements for as long as needed before being recovered. In the military world, large numbers of air-dropped gliders would be useful for locating submarines.

Navy researchers are looking at the possibility of gliders that can expel water and take flight again, something that has not yet been achieved. This would give them almost unlimited mobility and endurance, especially if fitted with solar cells. Researchers are also looking at networking gliders together via acoustic communications so they form an underwater sensing net, opening a large and permanent window into the underwater world.

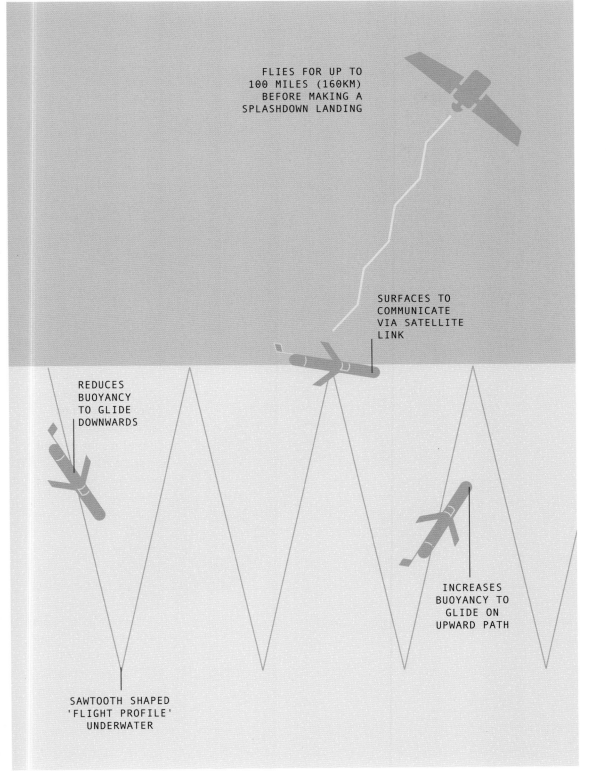

FLIES FOR UP TO
100 MILES (160KM)
BEFORE MAKING A
SPLASHDOWN LANDING

SURFACES TO
COMMUNICATE
VIA SATELLITE
LINK

REDUCES
BUOYANCY
TO GLIDE
DOWNWARDS

INCREASES
BUOYANCY TO
GLIDE ON
UPWARD PATH

SAWTOOTH SHAPED
'FLIGHT PROFILE'
UNDERWATER

KILOBOT SWARM CAN
ASSUME ANY SHAPE
REQUIRED

KILOBOT

Height	3.4cm (1.3in)
Weight	4g (0.15oz)
Year	2015
Construction material	Metal
Main processor	ATmega 328P (8bit @ 8MHz)
Power source	Battery

The Kilobot is a battery-powered widget the size of a table-tennis ball that moves at about a centimetre a second by vibrating its four legs. It may not look impressive, but an individual Kilobot is really a building block, part of a bigger entity. Kilobots become interesting when they swarm.

There are plenty of different types of insect swarm in the natural world. Bees forage efficiently over a wide area, with successful individuals sharing their finds with others. Desert ants work together to carry loads fifty times their own weight. Termites just a few millimetres long cooperate to build giant mounds metres high with complex air-circulation systems. In each of these cases, the individual carries out relatively simple actions, but the net result is 'intelligent' behaviour by the group as a whole.

Small, cheap robots using a swarming approach could combine for tasks impossible for larger robots. Swarms can cover a large area. They are robust; even if some members break down, the swarm can still function, whereas one broken motor or loose cable can bring a single robot to a halt.

Swarming approaches are well established in software, with some search engines using swarms of 'software agents' that spread out to locate data. Swarming robots are harder to develop than software, partly because of the cost. The other problem is complexity of dealing with so many machines at the same time. Every robot in a swarm needs to be charged up and switched on, and the appropriate software loaded onto it. The more robots, the longer this takes, and laboratory swarms are typically limited to a few tens of machines.

Harvard researchers, led by Michael Rubenstein, wanted to work with much larger swarms, and so developed an affordable robot designed for use in large numbers. Hence Kilobot, the kilo- prefix meaning one thousand. Kilobot is powered by a small lithium button battery. This drives the vibrating legs, which have a forward speed of about 1cm (0.4in) per second, and it can make a ninety-degree turn in two seconds. The Kilobot has a basic commercial processor for a brain, an infrared transmitter/receiver for communication and sensing, LEDs for signalling and a light sensor. The parts for one cost just £10, and it can be assembled in about five minutes.

Once assembled, Kilobots are 'scalable'; they do not require individual attention and ten thousand can be managed as easily as ten. The whole collective can be turned on or off remotely via an infrared communicator. The same method is used to download new software. To recharge, the robots are placed en masse between two charging plates.

The key to collective behaviour is having a set of rules followed by all members of a swarm. The behaviour of each member affects all the others – for example, when they are trying to spread out or cluster together. This can create complex interactions so it is important to test with real hardware rather than computer models to find out if it works in real life.

A Kilobot collective can follow a path, forming itself into a specific shape, or spread out evenly to cover an area, filling in any gaps. These simple tasks are the basis of what a robot swarm will do in the field, whether they are agricultural droids looking for weeds, or military robots on patrol. The Kilobot gives researchers a convenient testbed for software – for example, processes for splitting and reforming swarms – rather than having to jump straight in with something costing millions.

Swarming robots are already starting to appear outside the laboratory. Intel's Shooting Star quadcopters are an alternative to fireworks, producing colourful aerial displays of flashing LEDs; Lady Gaga performed with four hundred swarming drones at the Super Bowl halftime show in 2017. Meanwhile, the US military are testing swarms of small Perdix air-launched drones that could overwhelm defences.

Swarms of small robots may take over tasks such as mowing lawns and cleaning windows carried out by larger machines, and tiny swarming medical robots may one day carry out operations inside a patient. That is why it is important to have a good test platform like Kilobot to ensure the basic software works reliably.

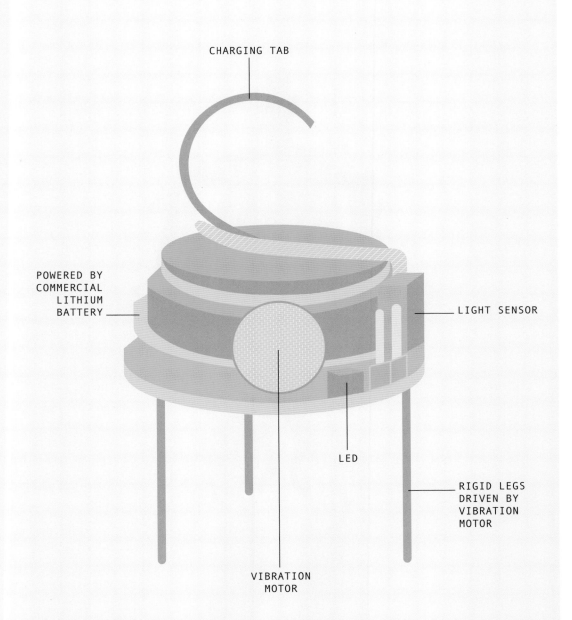

CHARGING TAB

POWERED BY
COMMERCIAL
LITHIUM
BATTERY

LIGHT SENSOR

LED

RIGID LEGS
DRIVEN BY
VIBRATION
MOTOR

VIBRATION
MOTOR

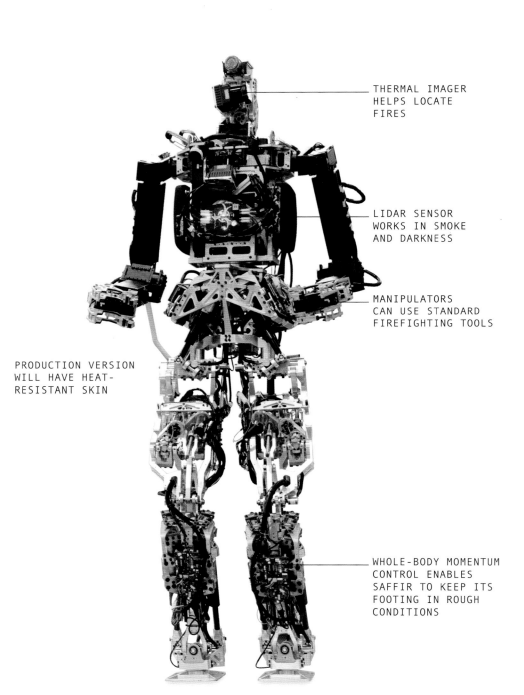

THERMAL IMAGER
HELPS LOCATE
FIRES

LIDAR SENSOR
WORKS IN SMOKE
AND DARKNESS

MANIPULATORS
CAN USE STANDARD
FIREFIGHTING TOOLS

PRODUCTION VERSION
WILL HAVE HEAT-
RESISTANT SKIN

WHOLE-BODY MOMENTUM
CONTROL ENABLES
SAFFIR TO KEEP ITS
FOOTING IN ROUGH
CONDITIONS

1.75M (5.75FT)

SAFFIR

Height	1.75m (5.75ft)
Weight	64kg (141lb)
Year	2015
Construction material	Steel
Main processor	Not known
Power source	External power

'Damage control' is an innocuous phrase describing what a warship's crew does in order to keep it afloat and in operational condition after it has been hit. The reality is a chaos of noise, smoke and torrents of seawater. Flooding is a hazard, but fire is more dangerous. Ships are loaded with flammable fuel and ammunition, and any fire can be catastrophic. Of the five US aircraft carriers lost during the Second World War, only one sank, the other four succumbed to out-of-control fires. To help prevent this in future, ships could have a new crew member. Step forward the Shipboard Autonomous Firefighting Robot, or SAFFiR, developed by the US Office of Naval Research, will assist with damage control.

The prototype looks like a classic mechanical man. Like Atlas and Robonaut (see pages 171–3 and 183–5), it is a two-legged humanoid, designed to work in human spaces and with human tools. It is able to open and close doors and hatches, and use existing firefighting gear.

SAFFiR's limbs have twenty-four degrees of freedom, and because it has to work even in stormy conditions, SAFFiR needs sea legs, or what the developers call 'whole-body momentum control'. This is a control system that moves all of the robot's joints together to keep its centre of mass supported on uncertain and unstable surfaces. Humans naturally take short steps and use their arms for support without thinking, but this is a novelty for robots.

The inside of a burning ship is an extremely difficult environment for firefighters. Thick smoke makes it impossible to breathe or to see anything. There is no question of staying well back and playing a firehose from tens of metres away as firefighters can on land. As it does not breathe, SAFFiR is unaffected by choking or toxic fumes. The prototype does not have thermal shielding, but the finished SAFFiR will be able to withstand higher temperatures for longer periods than human firefighters. It will also be waterproof and may function underwater in flooded sections of the ship. And, unlike human firefighters, SAFFiR is expendable. If someone has to go into an inferno on a one-way mission to close a hatch and stop flooding, a robot is the best candidate.

The robot has two sensing systems, a thermal imager and LIDAR (see page 44), both able to see through smoke. The thermal imager can locate and assess fires. Being able to see temperatures allows SAFFiR to detect a fire on the other side of a hatch or bulkhead. Its sophisticated thermal-vision system distinguishes the texture and motion of heat sources, so that SAFFiR can tell actual fire from hot material, and can separate heat reflections from heat sources. It can pinpoint the source of the fire and direct an extinguisher at it.

While the prototype SAFFiR is still remote-controlled, ultimately the robot will be autonomous, able to navigate through the ship, clambering over obstacles on its own. It will work as part of a team alongside human firefighters, taking orders verbally or by gestures, or even by touch. Navy research found that in damage-control situations, crews communicate by pushing or pulling each other to warn of danger or the need to move.

SAFFiR will not be the only robot on the team. The same project is also developing microflyers, small quadcopters that can negotiate the narrow confines of a ship at high speed and beam back sensor data. These will help assess damage and direct SAFFiRs to where they are most needed.

SAFFiR will be very much at home inside a fire, and a land-based version is a natural extension. But such a capable piece of kit should not sit idle in a locker, waiting for an emergency. Developers are looking at ways for SAFFiR to carry out routine shipboard tasks. If decks are swabbed and rails polished in the 21st century, SAFFiR's relatives will be doing the work. But this robot's real value will not be revealed until the ship becomes an inferno and more than human toughness and coolness are required.

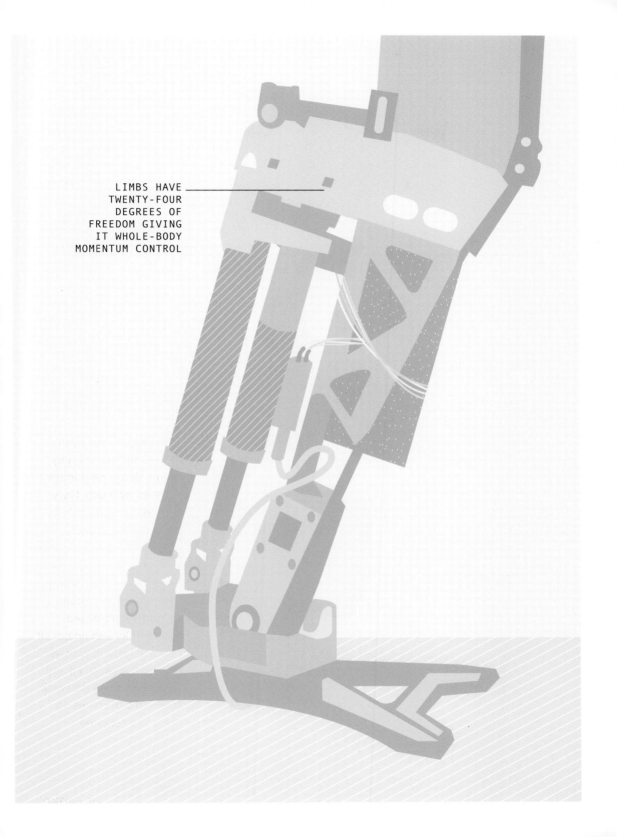

LIMBS HAVE
TWENTY-FOUR
DEGREES OF
FREEDOM GIVING
IT WHOLE-BODY
MOMENTUM CONTROL

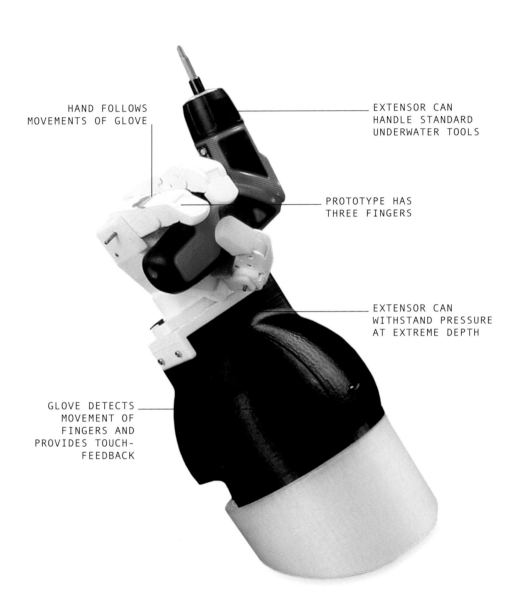

HAND FOLLOWS
MOVEMENTS OF GLOVE

EXTENSOR CAN
HANDLE STANDARD
UNDERWATER TOOLS

PROTOTYPE HAS
THREE FINGERS

EXTENSOR CAN
WITHSTAND PRESSURE
AT EXTREME DEPTH

GLOVE DETECTS
MOVEMENT OF
FINGERS AND
PROVIDES TOUCH-
FEEDBACK

45CM (17.7IN)

VISHWA EXTENSOR

Length	45cm (17.7in)
Weight	1.8kg (4lb)
Year	2014
Construction material	Titanium
Main processor	Commercial processors
Power source	External power

The human hand is a uniquely capable manipulator, reproducing its capability is hugely difficult. The challenge is even tougher in deep-sea environments and other similar extremes where handling often needs to take place at a distance.

Scuba divers can descend to about 100m (328ft) at most, beyond that the pressures become too great for humans to bear. For deep dives, divers are encased in an Atmospheric Dive Suit, a hard shell like a human-shaped submarine containing a bubble of air at normal atmospheric pressure. The pressure and thickness of the suit make it impossible to use gloves. Instead the suit has 'prehensors' resembling lobster claws. These are crude tools, with only open and closed positions – a single degree of freedom, in robot terms – making it difficult to grip irregular objects. It is impossible to use normal tools with prehensors, and even the most basic tasks, such as unscrewing a bolt, can be time-consuming and frustrating.

This problem inspired Bhargav Gajjar, CEO of Vishwa Robotics in Cambridge, Massachusetts, to develop a new type of remote-controlled robotic hand. The operator puts their hand inside a glove, and the robot hand moves according to the movements of the glove. Sensors in the hand provide force feedback, so the operator feels the resistance that the robot hand is meeting. This touch sensation, called 'haptic sense', allows the operator to handle objects as gently or as firmly as they need to.

The prototype Vishwa Extensor had three fingers and an opposing thumb – Gajjarinitially decided against giving it four fingers, as the ring finger and little finger tend to move together as one - later versions have four fingers for added strength. Each finger has four degrees of freedom, and the wrist has three, so it can follow the movements of the operator. Like the bebionic hand, it is the realization of Leonardo's dream of a machine that matches human articulation.

The Extensor can handle all the normal hand tools used by deep-sea divers. It is easy for this robot to use a wrench, pick up a nut and attach it to a bolt, or hold an electric drill and operate the trigger, all tasks that are difficult, or impossible, with claw-like prehensors. Opening submarine hatches, a crucial task, is child's play with the Extensor.

Gajjar developed the hand for US Navy divers and unmanned submarines. A diver would use the hand to manipulate objects at close range, or an operator on a ship could use a robot fitted with hands to work remotely with objects on the seabed.

The military are likely to use the Extensor for mine disposal, salvage and crash retrievals, as well as installing and maintaining underwater infrastructure such as sensors and communication cables. Nonmilitary users are also likely to appreciate the new possibilities offered by the Extensor. Marine biology and archaeology both require very careful handling of delicate structures – for example, when picking up undersea creatures like corals – for which the Extensor will be useful. In commercial diving, the Extensor allows divers to utilise tools such as chipping hammers and welding torches, which has been difficult with prehensors.

In future, Extensor may find applications far removed from diving. Gajjar suggests that it will be equally valuable for the remote handling of dangerous materials in laboratories or industrial settings – radioactive substances or hazardous biological samples. In space, this type of extensor would turn a machine like NASA's Robonaut into a remote-controlled pair of hands that could be used from inside the ISS, removing the need for astronauts to go outside into danger for jobs too tricky for the robot.

There is no reason for an Extensor to be the size of a human hand. A scaled-down pair of hands that a surgeon can position inside a patients' body to carry out an operation might be useful for surgery. These would be more dexterous than the surgical tools used by the da Vinci Surgical System (see pages 67–9). Or the extensor might be scaled up to giant size, with strength to match, for underwater or space work when dealing with large jobs. The only limits will be the imagination of the users.

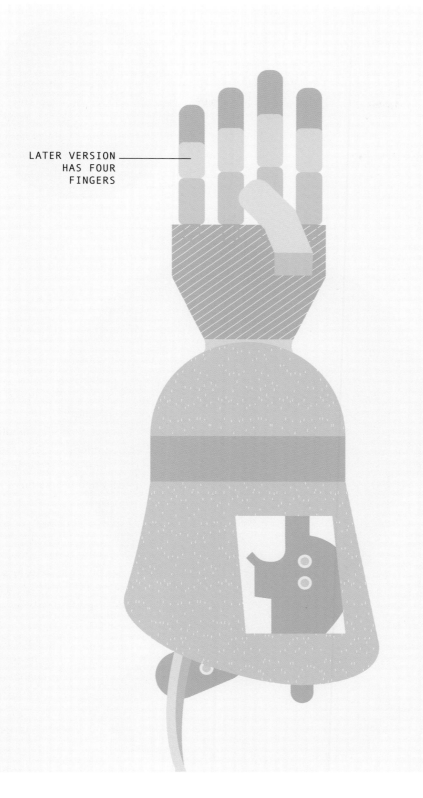

LATER VERSION
HAS FOUR
FINGERS

THE GEMINOID HI-4 IS A COPY OF ITS CREATOR, PROFESSOR HIROSHI ISHIGURO

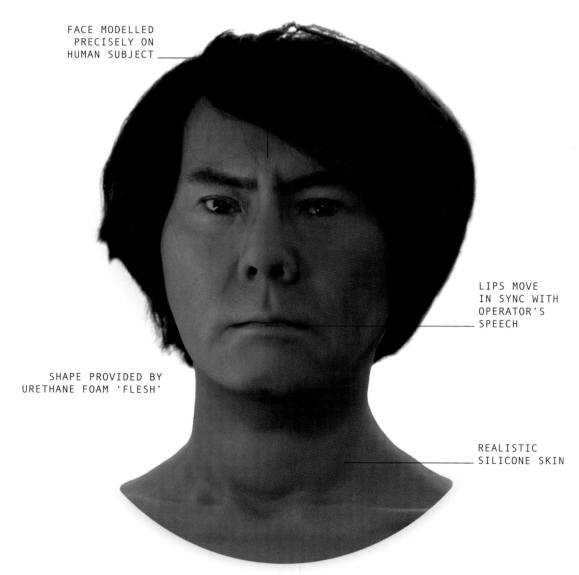

FACE MODELLED PRECISELY ON HUMAN SUBJECT

LIPS MOVE IN SYNC WITH OPERATOR'S SPEECH

SHAPE PROVIDED BY URETHANE FOAM 'FLESH'

REALISTIC SILICONE SKIN

GEMINOID HI-4

Height	1.4m (4.6ft) sitting
Weight	20kg (44lb) estimated
Year	2013
Construction material	Metal skeleton
Main processor	None (remote control)
Power source	External electric power and pneumatic air

Geminoids are eerily accurate robot copies of living people, built to explore the ways in which robots and humans interact. That eeriness is a key part of what they are about.

In 1970, Japanese robotics professor Masahiro Mori proposed the idea of the 'uncanny valley', a dramatic dip in the acceptance of robots as they gradually become more human-like. Mori noted that people are happy to accept industrial robots and other machines that look like machines, and even happier with androids that appear perfectly human. But in between these two peaks of acceptance is a dip, what he called uncanny valley. Anything that is not quite human tends to create strong revulsion.

Makers of dolls and puppets have always been wary of this effect. They know that to be likeable, their creations should be cartoonish rather than strictly realistic. Mannequins in shop windows are fine when they are abstract or stylized, but when they get too close to being human, the lifeless, staring faces start to look creepy. Animatronics can be highly realistic, but, as Mori observed, there is a moment when we suddenly see that they are not human, and that can trigger a plunge into uncanny valley.

Uncanny valley is a fundamental problem for designers and makers of 'social robots' that are meant to interact with people, especially robots that care for the elderly or dealing with children. Developers want their robots to be human-like so they can communicate in friendly, relatable and human ways without accidentally making people feel uncomfortable and possibly freaking them out.

Hiroshi Ishiguro, a professor at the School of Engineering Science at Osaka University, is at the forefront of research in this area. Ishiguro introduced the idea of the Geminoid, a robot that is not simply like a human, but that is an exact replica of a specific person. The Geminoid HI-4 is a copy of Ishiguro himself. In one sense, the HI-4 Geminoid is a cheat. While it may appear to be human-like, the HI-4 is a remote-controlled device with no intelligence of its own. Sixteen pneumatic actuators, twelve in the head and four in the body, allow it to mimic facial expressions and movements of the operator via a telepresence rig.

One of the aims of the HI-4 is to explore the human presence, and how we derive the idea of someone being in the room with us. The limitations of AI meant that remote control was the easiest means of studying the effect in the laboratory and providing a testbed for human-robotic interaction.

The HI-4's sculpted urethane foam flesh and silicone skin give a close approximation to the original. Ishiguro says that the HI-4 provides a genuine sense of his being present, and people who know him react to the robot as they do to him. The effect works both ways; when people talk to the robot, he feels they are talking to him directly even though the experience is remote.

Geminoids are being used to directly explore uncanny valley. Following on from Mori's original paper, researchers are looking at two key factors, eeriness and likeability, which can change after repeated encounters with the robot. Likeability was influenced by the way the robot behaved rather than anything else, and was not affected by exposure. However, repeated exposure to the Geminoid did gradually reduce the sense of eeriness. As with other new pieces of technology, an android might seem strange at first, but becomes part of the furniture after continued contact.

In one set of tests, around forty per cent of people who had a conversation with the Geminoid reported an uncanny feeling, with twenty-nine per cent enjoying the conversation. The results from these tests, and the particular aspects of the robot that people picked up on as being uncanny – bodily movements, it's facial expressions and the way it directed its gaze – are being fed into the design of the next generation of robots.

Whether it will ever be possible to build a robot that will pass as human remains to be seen. From the researchers' point of view, it would be a big enough prize to come up with a design that, while not necessarily human, is in no way uncanny, and that everyone is happy to talk to.

SIXTEEN PNEUMATIC
ACTUATORS ALLOW IT
TO MIMIC THE FACIAL
EXPRESSIONS AND
MOVEMENTS OF THE
OPERATOR

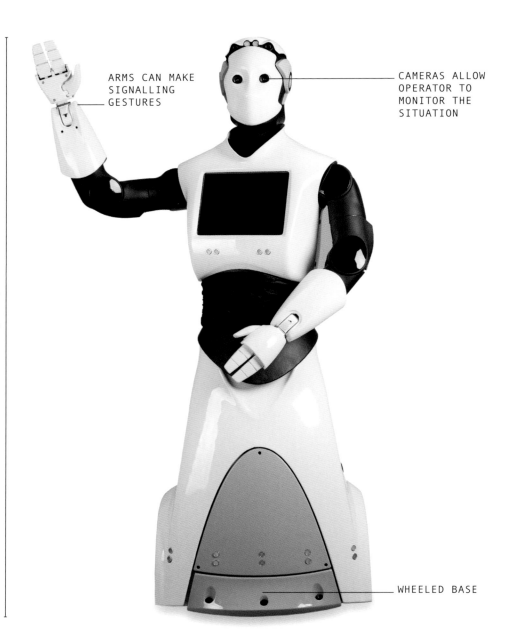

1.7M (5.6FT)

ARMS CAN MAKE
SIGNALLING
GESTURES

CAMERAS ALLOW
OPERATOR TO
MONITOR THE
SITUATION

WHEELED BASE

REEM DUBAI POLICE

Height	1.7m (5.6ft)
Weight	100kg (220lb) estimated
Year	2017
Construction material	Composite
Main processor	Commercial processors
Power source	Battery

The 1989 movie Robocop made a huge impression on the public psyche. The title character was a cyborg, part human and part machine, who tackled armed gangs using massive firepower. Robocop was an effective enforcer, but not your friendly neighbourhood bobby. This has not put off the government of Dubai from calling its own police droid Robocop.

Dubai's Robocop is nothing like the armed and armoured movie character; based on an existing robot made by Spanish company Pal Robotics, it is more like a mobile information kiosk. It is derived from the company's general-purpose machine designed for exhibitions and conferences, with a touch-screen communication panel on its chest.

The idea is that people will be able to report crimes, get information and pay traffic fines. The Dubai police are reportedly working with IBM's Watson technology so that Robocop can carry out a conversation, for example interviewing a witness and asking intelligent questions rather than simply following a set script.

It is doubtful how much of Robocop's workload really requires a physical presence. A decent website or mobile phone application could provide the same crime-reporting and fine-paying facilities. It could also talk to people and interview them wherever they were, and deal with any number of conversations at the same time. It would be accessible to everyone with a smartphone rather than being limited to just one location.

However, Robocop's main purpose seems to be public relations. By being seen in shopping malls and public places, it will get people used to the idea of robotic policing. Specialist police units have used robots for bomb disposal for decades, and there are tentative moves towards establishing police drone units, but these have met considerable public resistance. People do not quite trust robots, a fear that was reflected and amplified by the original Robocop movie.

Hence the need for a gentle introduction, in the form of the REEM robot. As well as acting as an information point, it will patrol the streets, using cameras as mobile CCTV. It can watch, record and arrest criminals, but cannot detain or pursue them – especially if they use stairs as a getaway.

Dubai wants one-quarter of its police to be robots by 2030. Robocops cannot replace humans on a one-for-one basis, but there will be other types of robot, too. In June 2017, a few weeks after announcing Robocop, the Dubai government showed off a miniature self-driving police car. This is the O-R3, an electric vehicle developed by a Singaporean company OTSWA Digital, equipped with 360-degree cameras to keep an eye out for crime with automatic 'anomaly detection' software. It can even launch a quadcopter to follow suspects.

It is a short step from this technology to machines that can subdue suspects. Quadcopters equipped with Tasers, pepper spray and other weapons have already been tested. Fielding them has more to do with political will and what is acceptable than with the limits of robotics. Some suspect a sinister motive for pushing this controversial technology; a core of robot officers, breaking up demonstrations or arresting protesters would be a mechanical process. Robot police would be ideal for a police state.

The issue is not the robots themselves but how they are used. When dealing with a potential bomb, police send in a robot from a safe distance. It makes just as much sense for a robot to tackle a potentially dangerous suspect. Providing two-way communication to an officer via microphone and speaker, and with no threat to their own safety the officer could respond coolly and rationally. At worst, a drone or robot would be damaged.

Patrolling robots could make sure that there is always a police presence nearby, and could extend the capacity of overstretched forces. With built-in cameras, everything robot police do is recorded and could be used in evidence. They cannot turn a blind eye, and are as impartial as justice is supposed to be.

The use of Robocops needs to be monitored closely, but we should not ignore their possibilities and benefits.

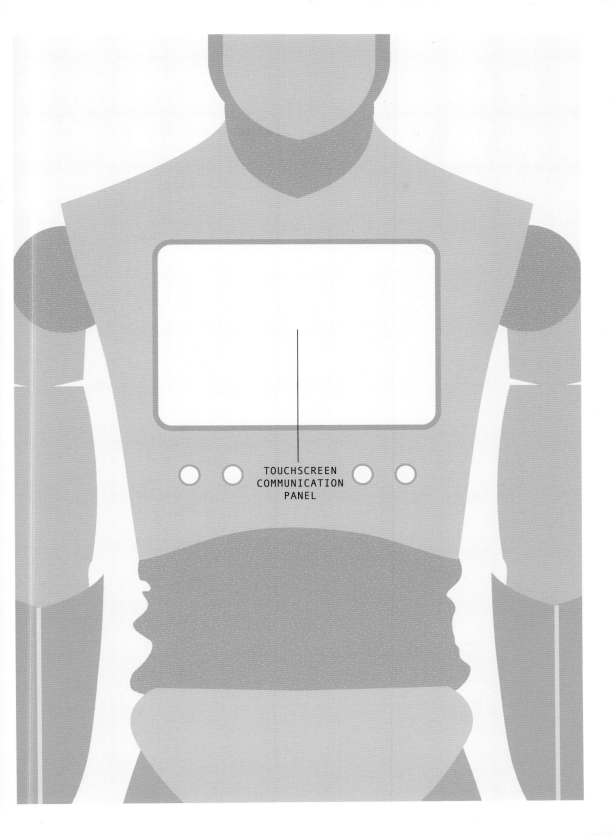

TOUCHSCREEN
COMMUNICATION
PANEL

David Hambling is a technology journalist and author based in South London. He writes for the *Economist*, *New Scientist*, *WIRED*, *Aviation Week*, *Popular Mechanics* and *Popular Science* among others. His first book *Weapons Grade* looked at the surprising military roots of modern technology, and his second *Swarm Troopers* examined the rise and future prospects of drone warfare.

Liron Gilenberg is a freelance graphic designer and illustrator specialising in books. Working with a distinct minimalist and meticulous style, she spends her time creating book covers, illustrations and layouts for clients such as Bloomsbury, Thames & Hudson, Kogan Page, and Quarto. Liron also runs workshops for publishers on briefing, design processes and presentation skills. A selection of her work can be seen on www.ironicitalics.com.

PHOTOGRAPHY

p.12 ©FANUC Europe Corporation; p. 19 © Agrobot; p.20 © Serbot AG; p.24 © Lely International; p.31 Photo by Stephen Brashear/Getty Images; p.32 © Australian Centre for Field Robotics; p.36 © Universal Robots; p.43 Photo by Thierry Falise/LightRocket via Getty Images; p.44 © Pure Technologies; p.51 © Jackie Shumaker for Epiroc Drilling Solutions, All Rights Reserved; p.52 Image of base: courtesy of Airobotics; image of drone: photographer: Rina Castelnuovo/Bloomberg via Getty Images; p.59 Images of machine components: ©David Kover for Serious Eats. Image of burger on conveyer: Momentum Machines/University of Pennsylvania; p.64 © irobot; p.68 © 2018 Intuitive Surgical, Inc. Used with permission; p.72 Photo by Neil Godwin/T3 Magazine via Getty Images; p.76 Photo by Ethan Miller/Getty Image; p.80 © Sundry Photography / Shutterstock.com; p.84 © Fraunhofer IPA, Photographer: Rainer Bez (2015); p.88 © Medrobotics; p.95 © Amazon; p.96 © by Ottobock; p.100 © DJI; p.104 © Wyss Institute at Harvard University; p.108 © Azad Shademan, Ryan S. Decker, Justin D. Opfermann, Simon Leonard, Axel Krieger and Peter C. W. Kim; p.112 © Vahana, A³ by Airbus; p.120 © Endeavor Robotics, PackBot; p.124 Photo by John B. Carnett/Bonnier Corp. via Getty Images; p.128 Courtesy U.S. Air Force photo/Paul Ridgeway; p.132 Photo by KIM DONG-JOO/AFP/Getty Images; p.136 © Boeing / Boeing Images; p.140 Photo by John B. Carnett/Bonnier Corporation via Getty Images; p.144 © Ghost Robotics; p.151 © U.S. Defense Advanced Research Projects Agency; p.152 Photo by Artyom Korotayev\TASS via Getty Images; p.159 Courtesy AeroVironment, Inc www.avinc.com; p.163 © Lockheed Martin; p.167 © XM Collection / Alamy Stock Photo; p.172 © HO/AFP/Getty Images; p.176 © Junzhi Yu/Chinese Academy of Science; p.180 © Hydronalix; p.184 © U.S. Defense Advanced Research Projects Agency; p.188 Courtesy NASA/JPL-Caltech; p.192 © Lori K. Sanders, Ryan L. Truby, Michael Wehner, Robert J. Wood, Jennifer A. Lewis / Harvard University; p.196 © Hanson Robotics Limited; p.200 Above: IHS Markit/Geoff Fein; below: © Dan Edwards, US Naval Research Laboratory; p.204 © Michael Rubenstein / SSR Harvard; p.208 Photo: U.S. Naval Research Laboratory/Jamie Hartman; p.212 Human like Deep sea Robotic Hand called "The Extensor" developed by Dr. Bhargav Gajjar of Vishwa Robotics and Massachusetts institute of Technology (M.I.T), sponsored by Office of Naval Research (ONR) and US Navy for Atmospheric Diving Suits (ADS) and Remotely Operated Vehicles (ROV). Image courtesy of Dr. Bhargav Gajjar of Vishwa Robotics and Massachusetts institute of Technology (M.I.T); p.216 Geminoid™ HI-4 has been developed by Osaka University; Geminoid™ HI-4: Osaka University; p.220 © Pal Robotics, 2017